Praise for *Shrinking the Technosphere*

Dmitry Orlov has written a clear and compelling
exploration of what is wrong with the technosphere,
and what we can do about it. This book needs to be read and
understood by policy-makers as well as the rest of us. It is
a valuable contribution to the resistance to the sacrifice
of the living planet on the altar of the machine.

DERRICK JENSEN, author,
Endgame and *The Myth of Human Supremacy*

The religion of technological progress cedes control
of our lives to machines and money. Dmitry Orlov tells us
how to return human values, pleasures, and freedoms to
the driver's seat. *Shrinking the Technosphere* is part self-help
book, part philosophical tour de force. It is both
entertaining and shockingly eye-opening;
it is a book that liberates the mind.

RICHARD HEINBERG, Senior Fellow,
Post Carbon Institute

A brilliant new book on a crucially important theme.
Our dignity, our autonomy, and quite possibly the survival of
our species depends on our willingness to extract ourselves from
the dysfunctional and metastatic mess that modern technology
has become, and craft a new relationship with technology
and the world. *Shrinking the Technosphere* marks an
important step in that necessary direction.

JOHN MICHAEL GREER, author, *After Progress*
and *Dark Age America*

This book is simply essential reading. It will jolt you
out of your comfort zone, but do not let that put you off.
We absolutely need to take a critical look at our world and the
assumptions upon which our lives and society are based.
And we need to work out where we go from here, individually
and, more importantly, collectively. Dmitry Orlov guides us
through this process more effectively, and entertainingly,
than almost anyone else writing today.

NICOLE FOSS, Senior Editor,
The Automatic Earth

It was Ivan Illich who first described how our doctors
induce illness, our teachers dumb down our kids, our judges
institutionalize injustice, and our "defense" establishment
makes us insecure. Dmitry Orlov now tells us our most
beloved tools make us incompetent. Written with
delicious humor, this is an absolutely essential
guide to avoiding Revenge of the Idiots.

ALBERT BATES, author,
The Post-Petroleum Survival Guide and Cookbook,
The Biochar Solution, and *The Paris Agreement*

SHRINKING THE
TECHNOSPHERE

GETTING A GRIP ON
THE TECHNOLOGIES
THAT LIMIT OUR AUTONOMY,
SELF-SUFFICIENCY AND FREEDOM

DMITRY ORLOV

new society
PUBLISHERS

Cover design by Diane McIntosh. Cover image: © iStock

Printed in Canada. First printing November 2016

Inquiries regarding requests to reprint all or part of *Shrinking the Technosphere* should be addressed to New Society Publishers at the address below. To order directly from the publishers, please call toll-free (North America) 1-800-567-6772, or order online at www.newsociety.com.

Any other inquiries can be directed by mail to:

New Society Publishers
P.O. Box 189, Gabriola Island, BC VOR 1X0, Canada
(250) 247-9737

LIBRARY AND ARCHIVES CANADA CATALOGUING IN PUBLICATION

Orlov, Dmitry, author
    Shrinking the technosphere : getting a grip on the technologies that limit our autonomy, self-sufficiency and freedom I Dmitry Orlov.

Issued in print and electronic formats.
ISBN 978-0-86571-838-8 (paperback).--ISBN 978-1-55092-633-0 (ebook)

    1. Technology--Social aspects. I. Title.

TI4.5.O76 2016   303.48'3   C2016-906242-2
                               C2016-906243-0

Funded by the Government of Canada   Financé par le gouvernement du Canada   Canada

New Society Publishers' mission is to publish books that contribute in fundamental ways to building an ecologically sustainable and just society, and to do so with the least possible impact upon the environment, in a manner that models that vision.

www.newsociety.com

MIX
Paper from responsible sources
FSC
www.fsc.org   FSC® C016245

Certified
B Corporation

new society
PUBLISHERS
www.newsociety.com

CONTENTS

INTRODUCTION

OVER THE PAST two centuries we have witnessed a wholesale replacement of most of our earlier methods of conducting business and daily life with new, technologically advanced, more efficient methods. Gone are the old household chores such as stoking a cooking stove, churning butter, spinning and weaving, sewing, making pens out of goose quills and lining writing paper, and so on. Most people are happy with the high-tech replacements—microwave ovens, packaged food, cheap imported textiles and ubiquitous electronic devices that have relegated elegant handwriting to a quaint nonessential. We like being able to jet clear across the planet in less than a day on journeys that once took many months. We do not complain about the fact that local travel no longer requires harnessing a horse or two and that turning the ignition key is now all it takes to put the power of hundreds of horses at our disposal.

But all of these comforts, conveniences and luxuries have their sinister side.

First, the question of what exactly is efficient about this new arrangement is hardly ever examined. If the new ways of doing things are so efficient, then we should all be leading relaxed, stress-free, enjoyable lives with lots of free time to devote to things like art, dance, poetry—pursuits once affordable only to the privileged few—not to mention taking frequent sabbaticals and retiring as soon as we feel that we've done enough. The fact that this is manifestly not the case (people are busier and more stressed-out than ever and are forced to wait to retire until ripe old age) should already have set off alarm bells: the new technology may be more efficient for some, but is it more efficient for *you*? Indeed, it turns out that it is more efficient in terms of primarily just one thing: corporate profits. Even by this measure of "efficiency" the new technology turns out to be defective if we take into account damage to the environment, the negative effects of this damage on us, and what it would cost to fully remedy them.

Second, the damage is not just to the environment but also to society. Although technological advances are always touted as "labor-saving"—because they boost productivity per unit of labor—many of them are, in fact, labor-destroying, because they don't merely enhance but *replace* human labor with machine labor, with the help of energy that is mainly derived from fossil fuels. A robot that replaces a human being does not boost that human being's productivity—it destroys it completely. Automation makes us economically superfluous. This would not be so bad if the robots worked for us, because then we could profit from them and devote most of our time to music, dance and poetry. But in a capitalist economy they belong to the capitalists who are few in number and, although the robots would work just fine without them, to them go all the spoils. The rest of us, once proud of what we could produce, are forced to work menial service jobs, until perhaps even these jobs come to be replaced by internet servers and yet more robots.

Third, although it is commonly thought that the machines work for us, this is increasingly not the case. Instead, more and more, it seems that it is we who work for the machines. We learn by taking online courses, where we please the machine by taking an automated quiz at the end of each unit. We faithfully listen to and follow phone mazes. We fill out numerous online forms. We squander our scant financial resources on endless technology replacements and upgrades, because technology is fragile and quick to become obsolete. Numerous technologists and troubleshooters, who are for the time being relatively secure in their employment, have to be on call 24/7 in case some bit of technology suddenly breaks. When it comes to our personal lives, there are dating websites to suggest mates for us, but are the matchmaking algorithms helping us find true love, or are they now simply breeding us like cattle?

Lastly, technology seems to be distorting our personalities. A century or two ago, nobody would ever say that people were "addicted" to their carpentry tools or their spinning wheel and loom. We may have loved our tools, lavished attention on them, kept them honed and oiled, decorated them with intricate paintwork and carvings, counted them among our most valuable possessions and proudly bequeathed them to our children. But they were mere useful objects—not fetishes—and they did not rule our passions.

Now, however, it is commonplace to hear of "internet addiction," and numerous sufferers seek medical treatment for it. More and more people are developing an unhealthy attachment to their smartphones: fondling them constantly; compulsively checking e-mails, Facebook status updates and tweets; and experiencing acute withdrawal symptoms the moment they lose network connectivity or the battery runs down. Back when horse and buggy was the preferred mode of transportation, people may have been fond of their horses but hardly thought of them as extensions or expressions of their personalities—as people now often think of their cars.

Now children grow up adept at video games but, because much of their experience of life is spent in various tiny artificial worlds which are manipulated using buttons and viewed through a pixelated screen, they grow up unable to discern or manipulate real objects in the physical world. Ask them to go dig up some potatoes or mend a fishing net, bake a loaf of bread or sharpen a pair of scissors—things that kids once grew up knowing how to do—and they would most likely scoff and tell you that they are not from some poor third-world country where people still have to do such things.

Technology deprives them of one of life's greatest pleasures: making things with their own hands.

Once our tools and machines were extensions of our bodies and minds, but now we are becoming slaves to our machines, dependent on them for our physical and psychological well-being and even our sense of self. Deprived of access to technology, we can no longer function and develop symptoms of anomie and depersonalization. In 2011 the UN declared that access to the internet is a human right and that disconnecting people from the internet should be regarded as a human rights violation. From a vantage point a few centuries back and, in all likelihood, a few centuries hence, this stance would seem as bizarre as declaring that it is a human right to inject heroin, or to ride unicorns.

As these four points indicate, the sinister side of even seemingly benign technologies is not particularly well hidden. Obvious symptoms of the technological ailments described above are easily observable all around us—if only we cared to look. Shouldn't they prompt us to question the assumption that technology is always helpful, useful and benign? But the prevailing, unquestioned belief is that technology is just wonderful, that newer technology is always better, that more technology is better than less and that, no matter what the problem, it is technology that will in the end save us. And don't tell us that we are dependent on it, that we let machines order us around, that we are mere appendages to machines, born to serve them until replaced, or that... heaven forbid... we are addicted to them! We can quit any time! (Right after compulsively poking at the smartphone one last time.)

And then there are all the technologies that are not the least bit benign: networks of machines that can exterminate all life on earth at the push of a button; technologies that monitor our every move, eavesdrop on our every conversation and attempt to predict our behavior so as to be ready to neutralize us even before we attempt to step out of line. These are technologies that we know and speculate about. But there are a couple more that, while commonplace, are not commonly regarded as technological frameworks, and although the notion seems exotic at first, this is indeed what they are.

We are now speaking of social machines, which control our thoughts and our behavior. They have some human moving parts (fewer and fewer every day), but they are nevertheless machines. They leave almost nothing for the exercise of individual free will and judgment—nothing that would counteract the simple imperatives of these machines to survive, multiply and amass power.

And then there are the political machines—engineered not just to produce certain election results but to give us the illusion of democratic participation and of having a voice in public affairs, while specifically depriving us of any meaningful choice and while simultaneously robbing us of our ability to think independently. And should these mind control methods ever fail, there is an ever-expanding technology suite that supports other methods of crowd control, including coercion, intimidation and the suppression of free speech.

Given all these negatives, it may seem appealing to turn away from technology altogether, smash all the gizmos and widgets and embrace the simple life of a hermit or a shepherd, or wander off into the woods and go feral or some such. But this book about technology is by a *technologist*, and the solution it offers is quite different. Instead of denigrating or repudiating technology, the idea is to wrest control of it.

To do so, we first have to learn to see it for what it is—by cutting through all of the buzzwords, the marketing hype, the pseudoscientific shibboleths and mumbo-jumbo. Then we have to learn to evaluate it: if it is efficient, then by what measure, and who stands to benefit from its efficiency? Efficiency as a euphemism for corporate profitability shouldn't fool us. Efficiency is a measure that relates productivity (output) to labor and resource inputs; it is meaningless unless we understand all the implications of these inputs and outputs. For a solar panel, does it simply input solar radiation and output electric current? No, its input is all the energy—mainly from fossil fuels—that went into mining, refining, fabricating, finance, design, research, sales, shipping, installation, tech support, maintenance and disposal. Its output is, yes, a modest amount of electricity. It could well turn out that your solar panel is a way to convert a lot of fossil fuel energy into a bit of electricity with the help of sunlight. How efficient is that? Perhaps it would be more efficient to use less electricity—or to not use electricity at all.

To avoid false efficiencies we have to learn how to choose our technologies. For any given technology, is it more efficient for us if the person who sells it to us sells more of it, or is it more efficient for us to buy less of it, need less money and not have to work as much?

Is any given piece of technology truly essential? If so, does it preserve our autonomy and freedom of action, or does it limit them in sneaky ways? Does it liberate us, or does it create patterns of dependence? Does it help us stay healthy, or does it contribute to mental or physical illness? Does it isolate us or throw us together with random strangers, or does it bring us closer to the people we like to spend time with?

Lastly, we have to learn how to optimize it: how will we get the most independence, free time, health and pleasure out of life using the technologies we do decide to use?

This is what the expression "shrinking the technosphere" really means: bringing technology down to a manageable number of carefully chosen, essential, well understood, reliable, controllable elements. It is about regaining the freedom to use technology for our own benefit and on our own terms.

Technology is always and everywhere bound up with the economy. While we cannot ignore economics altogether, we need to put it in its place, because a purely utilitarian, strictly by-the-numbers approach to all aspects of life is deeply flawed and altogether unsatisfactory if our lives are to have meaning. The economy isn't what matters—not the macroeconomic imperatives of economic development, growth, productivity or technological progress; nor the microeconomic imperatives of profitability, market share, innovation, brand loyalty or fashion. Rather, what matters is an economy of personal means, one which moves us away from being economic actors and toward being *economical* actors—economical and parsimonious in our use of technology. "Economic" to "economical": the change is slight, but it makes all the difference.

I
=

THE TECHNOSPHERE DEFINED

Its hapless denizens

PEOPLE WHO CURRENTLY inhabit any of the economically developed, industrialized parts of the planet have very little contact with nature. Most of their time is spent in climate-controlled environments sealed off from the elements. Bipedal locomotion—a hallmark human trait, alongside the opposable thumb—is decidedly out of favor. Now people move mostly on wheels, and when they do perambulate it is mostly across the parking lot or along supermarket aisles. When they do step off the pavement, the linoleum or the wall-to-wall carpeting, it is usually onto a well-marked "nature trail" from which they can observe nature without running the danger of actually touching any of it. Sealed off from nature, their bodies and minds are deprived of key natural inputs, and they develop a wide variety of ailments, from allergies and autoimmune disorders to autism and early-onset dementia.

Soon after birth they are injected with vaccines, saving them the trouble of maintaining genetic resistance against several common pathogens. When they later give birth, at the first sign of trouble during delivery they are offered cesarean section, saving them the trouble of maintaining compatibility between the outside diameter of the fetal cranium and the inside diameter of the

8

birth canal. When they catch an infection, they are treated with antibiotics (which are becoming less effective over time, as the bacteria evolve resistance against them faster than new antibiotics can be synthesized). Those who have significant medical problems are kept alive using a variety of aggressive medical treatments and allowed to reproduce, passing along their propensity for disease. Those who are incapable of reproducing naturally are offered fertility treatments.

While all of these measures can be said to improve the health of the population, they also eliminate the process of natural selection by which species maintain a healthy gene pool and evolve. The short-term result is better health and improved longevity; the long-term result is a polluted, depleted gene pool and a nonviable species. In the medium term, as numerous people around the world lose access to medical services because of unfolding economic failure, political upheaval and war, the results are already dire. Previously suppressed diseases reemerge. (The Ukraine is now becoming Europe's incubator for polio, which was eradicated when the Ukraine was part of the USSR.) Infant mortality surges. Women die in childbirth. People previously kept alive using insulin injections, dialysis, pacemakers and drug regimens all die.

All of these troubles stem from the fact that these people no longer directly inhabit the **biosphere**—the natural realm in which all life exists, and which provides us with breathable air, drinkable water, both wild and cultivated sources of food, construction materials for our shelter, natural fiber, fur and leather for our clothing and much else. Instead, they inhabit the **technosphere**, which is a parasitic entity that has grown up within the biosphere and is now busy destroying it. And the biggest problem of all is that many of these people, who in many places make up the vast majority of the population, have lost their ability to survive outside of the technosphere. They have become like the many breeds of domesticated animals—pets or livestock—that can no longer survive when released into the wild.

Pity the biosphere!

THE BIOSPHERE AND the technosphere can both be conceived of as living organisms—integral entities that consist of what to the human mind appears as an infinite number of parts interacting in an infinite number of ways. Just as we are unable to enumerate the number of different kinds of living organisms that make up the biosphere, or determine how they interact, so we are unable to see all the different artifacts—part numbers, stock keeping units, model numbers, versions, bills of materials—that have built our industrial civilization.

But it is clear to anyone who cares to look that the technosphere and the biosphere are distinct.

- The biosphere predates humans by billions of years and, unless humans manage to sterilize the planet as they go extinct, and barring some planet-destroying cosmic catastrophe, will continue without them for billions more.
- The biosphere can split into separate ecosystems without sustaining damage, and these separate ecosystems can just as easily recombine.
- The biosphere can progress when conditions are good and regress when they worsen. For example, as the oceans warm, acidify and become polluted, their population shifts from plankton, krill, corals and fish to bacteria, jellyfish and other more primitive organisms.
- Seen as James Lovelock's Gaia, the biosphere at all times seeks to preserve the homeostatic equilibrium of the planet in a way that supports a diversity of life forms.

The technosphere exhibits none of these features:

- The technosphere is a very recent phenomenon. It really only took off in the 18th century and was created and is being perpetuated by humans.

- The technosphere cannot be split into localized sub-technospheres without sustaining massive damage because of a large number of technical interdependencies. It treats technological isolationism and attempts at self-sufficiency as political problems to be fought tooth and nail.
- The technosphere can only progress, because as it progresses it as a matter of course destroys its previous ways of accomplishing things. (How many typewriter, adding machine and letterpress manufacturers are still around?)
- The technosphere is pursuing infinite growth on a finite planet, consuming nonrenewable resources at an ever-accelerating rate and destabilizing the global environment. As a whole, it is incapable of maintaining homeostatic equilibrium with its environment.

The biosphere and the technosphere are in opposition—a fight to the death. One of them will win, but which one? And what does this question mean for us?

In the beginning . . .

THE TECHNOSPHERE DIDN'T just pop into existence one day, fully formed. It evolved over time and is the culmination of a long-term evolutionary trend.

The genus Homo was from quite early on a tool-making genus. It all seems to have started with a certain hominid called *Homo habilis* that lived between 2.8 and 1.5 million years ago. It was an ape-like being that did not resemble modern humans, but it did make tools. *Habilis* is Latin for "handy" and the Handy Man got his name because his remains are often accompanied by primitive stone tools. This was an evolutionary breakthrough: no animal before then, and no non-hominid animal since, has been known to sit and methodically strike rocks with other rocks to give them a sharp edge and then put them to all sorts of uses.

From that time we hominids progressed, over the intervening millions of years, to use digging sticks to dig out edible tubers, to weave baskets and nets for catching fish, to fashion javelins to throw at game animals as we chased them down and to use a variety of other tools. The notable thing about these tools was how slowly their design progressed: thousands of years would go by with no apparent changes. Another notable thing was that these tools were made by people for their own uses: there were no specialized tool-makers. Making tools was a skill that parents taught to their children. Finally, people used tools to mediate their interactions with wild nature—not to dramatically alter nature to suit them, not to construct an alternate environment for themselves that is almost entirely separate from nature, but simply to get along a bit better while remaining part of nature.

It evolves!

AND THEN JUST a few thousand years ago a major change took place: progress in tool-making became much more rapid, taking centuries rather than millennia to achieve major breakthroughs like copper, then bronze, then iron tools, the wheel, pottery and much else. The new tools gave us enough power to start *transforming* nature to suit our needs. Axes and ploughs transformed forests into fields, scythes and sickles brought in the harvest, picks and shovels dug canals and wheeled carts transported the harvest to population centers which all these tools made possible for the first time.

All indications are that the people who embraced these innovations were sicker, less happy and less secure than those who preceded them. Their health suffered because their diet was now restricted to just a few staples instead of the wide array of wild fruits and vegetables they had evolved to eat. They were less happy because they now had to stay put rather than wander about; to fit into a coercive, repressive social scheme; to follow orders and to work much harder. And they were less secure because instead of the variable but more than sufficient bounty of wild nature, they had to rely on crops, which frequently failed, causing malnutrition and starvation.

So why did they go on living this way, instead of perceiving the error of their ways and reverting to the older, easier, healthier ways? One obvious answer that suggests itself is that they were not acting in their own interests; they were acting in the interests of the artificial, synthetic entity—a social machine—which they had inadvertently created. Although this entity could be said to serve human needs on some level, it clearly had its own interests to pursue: to survive, to expand and to control all that it could. To this end, this entity—at the time no more sentient than an amoeba but already with a rudimentary volition of its own—evolved certain traits that allowed it to eventually enslave the majority of the people.

It did so by elevating a minority (shamans, priests, kings, emperors) above the rest. This elite lived in conditions far more luxurious than what was possible before, while the rest lived in conditions that were quite a lot worse. While previous social groups enjoyed equality, including gender equality, the new social groups were stratified and hierarchical. They subjugated women and reduced a minority of the population to the status of slaves.

It did so by engendering dependency. Instead of parents teaching children to make all that was needed to live, as parents had done for many millennia, people now came to depend on specialists who constructed their tools, their shelter, their clothing and much more. Instead of being quite capable of defending themselves against wild animals and other humans, they now were forced to depend on professional guardians who enforced a monopoly on violence.

And it did so by carving up territory. Whereas previously human bands inhabited flexibly defined domains that shifted with the seasons and with climatic changes, following the migratory patterns of animals and patterns of nature generally, now they were organized into larger tribes with explicitly defined borders between them. As everyone was forced to settle in their assigned spot and to stay put there, an entire range of previously common migratory and nomadic lifestyles became marginalized and sometimes even outlawed.

Once the territory was carved up and people were split up into tribes, they had to defend that territory or risk losing it. And this gave us war. Whereas earlier humans found fighting to be completely stupid—there were few of them inhabiting a rather large planet, so why not just avoid those you don't like?—now people had to fight, or they stood to lose everything. While before the humans were few in number but free and, on the whole, happy and healthy, with the advent of settled modes of living and agriculture humans became enslaved, miserable, sickly—yet ever more numerous.

It overcomes
its natural limits

THIS WENT ON for several thousand years. Civilizations and empires rose and fell in steady progression. Fertile lands failed from overgrazing, soils became too saline from irrigation, sporadic minor shifts in climate disrupted systems of agriculture that became too specialized, and each of these could take down entire societies. Some districts became depopulated; for instance, the former bread basket of Mesopotamia is now mostly barren desert. Many districts became deforested as land was cleared for agriculture and then failed due to soil erosion. Others—Greece, for example—managed to survive by shifting from growing soil-depleting annuals such as wheat to growing and exporting the fruits of more sustainable perennials, such as wine made from grapes and olive oil from olives, and importing the wheat. A few others—most remarkably, Rome—depleted their native lands, but managed to eke out a few more centuries of supremacy by surviving on tribute.

But in all of this the technosphere was restricted in two major ways, which made the damage it caused self-limiting. The first restriction had to do with energy: all that was available to the technosphere came from sunlight captured by plants through photosynthesis and used directly as fuel or indirectly as muscle power, both animal and human. Some amount of energy was available from windmills and waterwheels, but overall the technosphere had to live within a strict energy budget, which was based on what living nature could provide. The second restriction was geographic: technology did not yet exist to navigate the entire planet, plundering resources from wherever they were to be found. Whenever a civilization collapsed, populations dwindled, fields were abandoned and reforested, and the Earth recovered.

Conquest of nature

EVENTUALLY THIS TOO changed, and the technosphere was able to shift from *transforming* nature to suit its purposes to the task of *replacing* it. At some point, starting in the 18th century, technologies were developed which gave it access to energy from dead nature, sequestered in the earth's crust from millions of years ago. Coal, then oil, natural gas and eventually uranium could generate far more energy, in far more concentrated form, than could be obtained directly through photosynthesis or from other renewable sources—while supplies lasted. The second restriction fell away too: with the advent of sailing ships that could circumnavigate the planet, natural resources could be fetched from virtually any spot on the entire planet—as long as there were still places on the planet left to exploit.

No longer constrained by what the biosphere could provide locally and perpetually, the technosphere could now grow exponentially, repeatedly doubling in size. Nor was it any longer constrained by what could be achieved by human labor: ever since the invention of the waterwheel and the windmill, human and animal labor have been progressively replaced with machine labor. With the invention of the steam engine, which was initially used to pump water out of coal mines, the process really took off. Farmers, who once walked behind their ploughs, today sit in the air-conditioned cabs of their tractors, pushing buttons, while the tractors are guided with the aid of satellites. Nor is the farmer's job safe for long: the latest agricultural technology suite includes the driverless tractor.

With the role of manual labor largely eliminated, it is now the turn of intellectual labor to become redundant, as computers replace most of the work that was previously done by people. Computer algorithms now carry out many functions that were previously performed manually by humans, be it filing documents, making travel itineraries, choosing investments or selecting your friends and mates (on social media). Where humans cannot be replaced by robots—for instance, in teaching young children— the technosphere's goal is to make humans function in a manner that is as robot-like as possible: teachers are evaluated based on how well their students do on standardized tests. But only certain things can be tested in this manner: namely, rote memorization and mechanical skills. Everything that cannot be measured and must be evaluated based on human judgment ends up thrown overboard, because teachers are now forced to reduce teaching hours in order to focus on coaching students to excel at standardized tests.

What has happened in education is a particularly egregious example, but there are many others. In every realm, in every form of human endeavor, the technosphere's goals appear to be the same: to technicalize the field to the greatest extent possible; that is, to reduce the role of subjective human judgment, to rule out the use of intuition in reaching decisions and to eliminate spontaneous behavior by forcing everyone to act in accordance with written procedures. And while before it would have been difficult to enforce such strict compliance, the systems of electronic surveillance that are now in place in almost every workplace, which record every keystroke and capture every physical movement on video, force everyone to self-police and self-censor out of well-founded fear.

Previously, many job functions, such as working on an assembly line performing the same movements all day long, were tedious and unfulfilling, but now that most industrial workers have been replaced by robots those workers serve no function at all. Or rather, they have one residual function: to consume. But this brings up an obvious question: why should the machines—or rather, the owners of the machines—continue to pay workers who have been made redundant? Answer: they shouldn't, and they won't.

The end result of this process is that countless millions of people have been made economically superfluous while, at the same time, countless people no longer have access to what they need to be self-sufficient. Here is a specific example: since practically forever, except in most extreme situations, everyone could count on being able to die in their own bed in the care of their families. But now that even death has been technologized and professionalized, they are virtually guaranteed to die in a hospital or a hospice, sporadically attended to by an underpaid minion. The only people who can still count on receiving good care on their death-bed are the very rich—not from their family members, mind you, but simply because they can afford hired help of a higher quality.

What they can count on is that the medical system will do all it can to artificially keep them alive—against their own wishes and often against the better judgment of their family members—and will sometimes even bring them back from the dead and keep them on life support in a persistent vegetative state (as happened to my own father). How likely this is to happen depends to a fair extent on whether they have sufficiently comprehensive health insurance, but it is not really a financial question. After all, lavishing resources on people whose prognosis is hopeless is not exactly economically advantageous. Rather, the question is one of blindly attacking the last vestige of nature in human nature—death, that is. In nature, death is an essential part of life; within the technosphere, death is a technical limitation to be overcome through technical means—hence all of the unhealthy focus on longevity, to the neglect of much else.

It wants to control absolutely everything

THE REASON FOR extending life for as long as possible, no matter how little sense this makes, is to be found in the abstract teleology of **total control**. The technosphere's compulsion is to control *everything*. It is unacceptable to it for old people to decide when to die all on their own. Death cannot be left up to a subjective judgment; it must be the objective outcome of a technical, measurable process. The idea of an old man, lying in his bed, saying good-bye to everyone, closing his eyes and drifting off is abhorrent to it. At the very least, you the patient (for you can no longer be old or sick without becoming one) must be attended by a specialist who is monitoring your pulse while looking at a watch in order to accurately record the time of your death. (Remember, accuracy is everything, in death especially!)

Nor is it acceptable for the family to wash, dress and lay out the body, to hammer together a coffin and to dig a grave in the back yard—no, it can't be that simple! While private burials are not altogether illegal, the bereaved are required to file a number of forms, including soil tests and hydrological surveys, and to obtain a number of permits. How long does that take? Also, the burial site has to be recorded on the deed for the property, and if the property is mortgaged or used as collateral on a loan, the mortgage-holder or the loan-holder can refuse to allow the burial to proceed. And if the land is sold, it may be necessary to exhume the body and to buy a burial plot at a cemetery—more permits and tens of thousands of dollars in expense.

And that would be the cheap option. What people normally do, when bereaved, is make a phone call to a funeral home. The loss of a spouse or a parent is a bad time to do comparison shopping, and so they are talked into spending an inordinate amount of money on burial expenses. But this option is only available to those who have money. As for the rest, in more and more cases the families have no choice but to have their terminally ill loved ones carted off to a hospital and then, to avoid the expense of the burial, they simply don't claim the body.

It wants to technologize everything

THE TECHNOSPHERE WANTS to make everything into a piece of technology. No profession is immune—not teaching, not caring for the sick and the dying. Creative professions, such as music and graphic arts—which are not technical because, although they make use of technique, they are not dictated by it—are nevertheless technologized. Music and images are digitized and distributed via the internet.

In the interest of exercising **total control**, access to everything must become technologically mediated. You can no longer make music with your hands and your mouth, or listen to music directly with your ears; you need microphones, mixing boards, amplifiers, or mp3 players and ear-buds. You cannot simply look at art; you need to download it and view it on a high-resolution screen. Creative professions that cannot be technologized—fine arts, dance, live performance arts, philosophy—are marginalized, or kept alive artificially to provide amusement for the rich. They are regarded as hobbies, and those who engage in them are not taken seriously unless they happen to be catering to rich connoisseurs and collectors.

The technosphere wants to control everything: judgment, initiative, intuition, spontaneity, autonomy, freedom—all of them subsumed under a mountain of written laws, rules, regulations and protocols. Humans are simply too unpredictable and too difficult to understand, and therefore the technosphere can't allow them to make their own decisions. The only ones entrusted with decision-making are the experts—the ones who can be relied upon to follow explicit procedures, for fear of losing their privileged positions.

And in order to control everything, the technosphere wants to *quantify* everything. Anything that could be significant—be it consumer preferences, sexual preferences, political opinions, religious beliefs, down to the average number of times an hour people blink or touch their faces—is to be measured and tabulated, because without numbers it is not possible to make objective decisions, and without objective decisions it is impossible to exercise total control.

It wants to put a monetary value on everything

SOME THINGS CAN'T be bought—or can they? From the point of view of the technosphere, if something is valuable, then it must, by definition, have a value, and value is quantified in terms of money. If you refuse to define something in terms of its monetary value, then, as far as the technosphere is concerned, that is tantamount to declaring that it is worth nothing.

What is the value of wild elephants? Well, they provide some profits for the safari business, the nature film industry and zoos. They also provide profits for ivory poachers, and while poaching is illegal, it does produce an undeniable boost to the global economy and, ipso facto, adds to the elephants' economic value. Animals that have no commercial potential—the vast majority of them—have no economic value at all.

From the point of view of the technosphere, the biosphere is simply there to provide it with resources and services. Its view of the biosphere demonstrates the technosphere's striking mental deficit: it is unable to see limits. Until it runs up against them, it simply can't see them and assumes that natural resources are infinite. And when it does run up against them, it invariably treats the problem as a financial problem.

For example, when oil prices spiked, it was automatically assumed that the problem had nothing to do with resource depletion but was entirely due to lack of investment in the oil industry. Sure enough, increased investment eventually resulted in increased production and a glutted oil market, but the fact that the increased investment became necessary had *everything* to do with resource depletion. What's more, the effect of increased investment is temporary; like rust, resource depletion never sleeps, and at some point the level of spending needed to maintain production will become impossibly high.

If something causes damage (industrial pollution, crime, soil erosion and so on), then its value is negative and calculated in terms of how much it costs to mitigate the damage. The technosphere seeks to optimize the amount of damage so that the cost of preventing damage is balanced with the cost of mitigating it. For example, as far as the technosphere is concerned, there should be an optimum amount of rape that happens in society, such that the cost of deterring rape, by prosecuting and incarcerating the rapists, does not outstrip what it would cost to mitigate the effects of rape by providing therapy for the victims.

When it comes to damage, the technosphere exhibits another quite striking mental deficit: it can't understand the idea that some things can't be fixed for any amount of money. It can't quite put a value on "ecosystem services" such as fresh water, breathable air, arable land and a survivable climate, without which life on Earth would not be possible. Consequently, it can't put a value on completely destroying them either—all because this level of damage can't possibly be fixed by throwing money at it.

Above all, the technosphere loves money because it makes possible every sort of technical manipulation. Money is the ultimate fungible commodity: any one dollar can be replaced by any other dollar. The same money that buys Girl Scout cookies can be used to buy shotgun shells to shoot up a Girl Scout cookie stand.

It is nearly impossible to do good by spending money, because whenever you spend money you inevitably feed this monster. It is not possible to oppose the technosphere using money because money is its own native medium. But it is possible to disrupt and block it by de-monetizing economic relationships using barter and gift, by disrupting the fungibility of money by using local scrip and chits and by making as much of the natural realm as possible off-limits to any sort of commercial exploitation.

It demands homogeneity

THAT THIS IS so is clearly visible in spite of all the talk about "diversity" and "multiculturalism" in the overdeveloped Western nations. When people say "diversity," what they really mean is homogeneity: a common, simplified, commercialized mass culture organized around nationalist/globalist concepts and symbols. In the pursuit of total homogeneity, "diversity" turns out to be quite useful. This is not a paradox but mere misdirection. Over time tight-knit communities tend to develop their own unique local cultures, traditions, languages and dialects, and these allow them to withstand the onslaught of outside influence. People who share a common local culture recognize and automatically trust one another. This is true diversity: the diversity of distinct, separate cultures with unique traditions of mutual aid, cooperation and solidarity. And it is this that makes them hard for the technosphere to dominate and to control.

And the best way to bring them to heel is to disrupt them by inundating them with floods of newcomers, ideally from entirely incompatible cultures, and to force them to deal with each other through administrative and police action assisted by the schools and the courts. A few generations later, once their children grow up together within a dumbed-down pseudo-culture that is designed to be inoffensive to anyone, all traces of local culture are gone except for a few vestiges of cuisine and dress. The local population is now denatured—a homogenous mob, part of a mass culture—and can easily be manipulated through mass media and advertising, having been rendered transparent, measurable, completely dependent and perfectly controllable. The fact that it no longer has any particularly compelling reason to exist is entirely beside the point. Those few who manage to hold on to fragments of their own culture are taught to regard it as their "cultural back-ground." And what then, pray tell, is their "cultural foreground"? It is a cultural wasteland.

Once local culture has been all but eliminated, great new vistas open up for homogenizing society, its institutions and the built-up environment. There is no longer any need for special arrangements to accommodate local languages, dialects, traditions or architec-tural vernaculars. The mix of commercial products offered to the communities can be made the same across the board, resulting in greater economies of scale. Once local architectural styles have been eliminated, it becomes possible to build cookie-cutter houses, the same everywhere, regardless of the natural environment or the climate.

Most importantly, the result is a great deal of social disruption, where the people, made rootless and no longer possessing a sense of loyalty or bearing an allegiance to any place or group, can be made to roam the culturally barren, homogenized landscape, flock-ing to where the jobs happen to be, then moving on once these jobs vanish. Masses of rootless humanity can be mobilized to work on projects which those who are rooted and invested in their com-munities would pass up. The recent shale oil boom in some of the western US states presented just such a spectacle. Entire makeshift towns sprang up and, when the boom predictably turned into a bust and jobs disappeared, they promptly turned into depopulated ghost towns filled with foreclosed and abandoned houses, while the human tumbleweeds for whom they were hastily hammered together moved on to look for work elsewhere, or turned to liquor, methamphetamine and synthetic opiates.

The ability to dislodge and then exploit people is a key ingredi-ent in the technosphere's success. In France, the ability to rapidly industrialize only opened up once the French Revolution had caused a great deal of social dislocation, disrupting pre-existing social relationships and local traditions. In England, industrializa-tion likewise was only able to proceed thanks to social disruption, when "improvements" in agriculture made much of the rural population superfluous. It then became possible to stop people from living off the land, as small landholdings were unified into large estates through the enclosure movement, which also co-opted land that was previously used as the commons. Waves of dispossessed peasants flooded into towns and, given no better choices, went to work in factories or were pressed into service as sailors, soldiers or colonial servants in the service of the empire.

A similar process of large-scale societal disruption took place in Russia after the Bolshevik Revolution of 1917 and after Mao Tse-tung's Cultural Revolution that spanned the decade from 1966 to 1976. In each case, long-standing social relationships based on local custom and tradition were blown away in a wave of political violence. In Russia, collectivization deprived much of the rural population of what had been a comfortable livelihood, forcing it into the cities to work in factories, rapidly transforming Russia from a predominantly agrarian nation into an industrial power-house and a world power that was able to dominate half the planet for several decades. In China, a similar wave of disruption swept through society, unleashing its industrial potential and converting it in a short span of time into the world's largest economy (as judged by the population's purchasing power) and the world's factory, where most of the world's manufacturing now takes place.

The goal of the technosphere is to reduce all of humanity to the level of a common herd, with a single, primitive language, a single, primitive, commercially controlled culture and a single system of unelected, technocratic governance. Luckily, this is quite unlikely to happen, since humanity happens to be comprised of several distinct civilizations that are unlikely to either merge or disappear. It is now becoming even less likely, since the Western model, dominated by Anglo and Zionist influences, is in increasingly ill repute throughout the rest of the world, which now includes well over half of the world's population, productive capacity and wealth. Nevertheless, it is important to see the drive toward a supposedly "multicultural" global commercial culture, global economic development and a harmonized, supposedly "democratic" system of global governance for what it really is: an attempt to erase just about everything that makes us human and that makes human life worthwhile.

It wants to dominate the biosphere

PARTS OF THE biosphere that the technosphere does not control it considers "wild": uncivilized, unruly, uncontrolled, unrestrained, unreasoned... unacceptable. Land that has not been deforested, plowed, bulldozed flat, paved, built upon or otherwise destroyed as far as the biosphere is concerned it calls "unimproved land." At the very least it has to be mapped (because everything must be measured) and inventoried to find out what natural resources are there to plunder.

All land, whether "improved" or in its natural, pristine state, must be owned by some entity that the technosphere recognizes—an individual, a transnational corporation or a national government. Under no circumstances can it be owned by the indigenous tribes found to be quietly living there as they always had; they are carted off to the nearest town, charged with vagrancy or trespassing, locked up, and eventually released and told to go and get a job. Should any of these people accidentally stray across a national border (which is an imaginary line that often runs along a ridge or a river) they are arrested and deported, and then the authorities on the other side of the border cart them off to the nearest town, lock them up and eventually tell them to go and get a job.

This is because nothing, and nobody, can be allowed to live outside of the technosphere and its control—humans especially. Wild animals can sometimes still get away with it, especially the ones who live in the ocean, but it is getting harder for them on land, as more and more of them find themselves tranquilized and their ears tagged—all for their own good, of course. Even migratory birds like Canada geese can't be left unmolested, and many of them can be seen walking around with a little yellow ring on their foot. When it comes to pets and livestock, all of these have to be microchipped.

And when it comes to humans, each and every one of them must be fully documented, and ideally each and every one would be fingerprinted and made to provide a DNA sample.

It would appear that the technosphere doesn't particularly like living things: they are all nonstandard, their behavior is hard to predict, and they are hard to control. They interbreed in funny ways and, worst of all, they don't stay the same, but evolve. Microorganisms are especially troublesome: as soon as the technosphere figures out how to control one of them, it starts trying to evade the controls by evolving. Also, microorganisms sometimes share their genes with other, unrelated microorganisms, not caring a whit whether the gene in question has been synthesized and patented by a biotechnology company. This is blatantly illegal, but how does one sue a microorganism?

The technosphere, for the time being at least, cannot survive without its human servants, who in turn need their animal and vegetable servants, and so the technosphere has to grudgingly put up with some amount of uncontrollable biological messiness. But it tries to minimize it. It is well-known that plants grow best as groups of different species. On a given patch of land, you can grow more corn, beans and squash when all are grown together than you can if each is grown separately. Forests grow best as a succession, starting with grass, then shrubs, then fast-growing, short-lived softwood trees, then slower-growing, longer-lived hardwood trees. But the technosphere detests all of this hard-to-measure, uncontrollable complexity and insists on agricultural and forestry practices that are based on monoculture: only one variety of one species is allowed; everything else is considered a weed and killed, biological efficiency and health of the ecosystem be damned.

Another way to minimize the uncontrollable messiness of living things is to classify every living thing as rigorously as possible. Animals can no longer reproduce as they wish but must be kept within specific breeds. This has been especially cruel with regard to dogs, but just about every domesticated species of animal, even the essentially untamable house cat, has been damaged in this way.

Breeds of hunting dogs are especially damaged—they are specifi-cally bred to be unable to hunt for themselves. Humans are being bred too: to sit behind a computer all day, to survive without much contact with any other living thing and to value numbers more highly than anything that's alive or even tangible.

Yet another major simplification, which the technosphere likes to make use of a lot, is to ignore biology and to concentrate on phys-ics and chemistry. Biology, you see, is something of a soft science. Living things are just too complicated. The biosphere abounds with species of bacteria, fungi, plants, microscopic animals, all of which interact in an unfathomable number of ways, and trying to map out their relationships is more like anthropology than a hard science. But there is a trick: you can just treat them all like bags of chemicals. And then things do get simpler. When viewed on a molecular level, it turns out that there are plenty of ways to manip-ulate them: you can splice their genes, you can alter their behavior using chemicals, and you can kill them. Killing them is especially popular.

If you think that the technosphere would limit its right to treat living things like bags of chemicals to lower life-forms, you are wrong. Humans are fair game, and not just their bodies. For a while now, the human brain has been the final frontier in this physical/chemical assault on living things. Electroconvulsive ther-apy—zapping the brain with electricity—has been popular since the 18th century. And now antidepressants—chemicals that alter brain chemistry in various ways—are being prescribed to a quar-ter of all women in the US (while the men seem to prefer large doses of alcohol). This is a little like trying to adjust an insanely complicated clockwork mechanism by poking it with a stick or by throwing handfuls of sand at it, but what alternative is there? It is quite well understood that individual madness is a symptom of a mad society, but since a mad society is all that there can be within the technosphere, fixing it is not an option. The only option seems to be to throw yet another handful of sand at it. And if it kills you... so what?

Perhaps the technosphere's most significant trait is that it generally doesn't care whether you live or die. It will try to make you happy, but failing that, it will just as easily kill you. It all depends on what turns out to be the most efficient within its grand scheme: It likes you if you are one if its obedient servants, preferably an engineer or a technologist. Scientists are valued too—but only those that do spectacular tricks that bolster the technosphere's prestige and sense of power. If you are not one of these favored specialists, then you can still eke out some sort of living, provided you produce and consume as you are told. If you refuse, then you can rot in jail, or be killed. It's all the same.

It controls you for its own purposes

THE TECHNOSPHERE DOESN'T particularly care whether you live or die, or whether you are happy or miserable. Its goal is to control you and to make you serve its purposes, which are to grow, to control everything and to dominate the biosphere. How it achieves this control is a matter of what is most efficient. If you are one of its faithful servants, then the best way to make you do your job well is to incentivize you—to give you high status, ample pay and lots of perks. But if you are a lowly menial grunt in its service who, unfortunately, cannot yet be replaced by a shiny new robot, then low pay and low status suffice, and destroying your autonomy and self-reliance while fostering your dependency is the key to making you perform. If you are a technologically useless person but harmless—an artist, a philosopher, writer, poet, free thinker—then the technosphere simply can't see you, because what you do is not measurable in units the technosphere can understand. But if your thinking turns out to be dangerous or harmful to the technosphere—because you are someone who tries to break the chains of dependency and to find ways to live outside of the technosphere, or to undermine it in some other way—then it will consider you as nothing less than a terrorist!

To the technosphere, morality is a purely functional concept: to it, the function of morality is to control humans, not to constrain machines. Religious doctrines are, likewise, to be co-opted as techniques for inducing herd behavior in large populations. If the population is socially liberal and decadent, the technosphere will champion gender equality and the rights of sexual minorities; if it is socially conservative and religious, it will instead help spread Christian fundamentalism or Islamic extremism. Politics, democratic and otherwise, is a smokescreen behind which lurk political machines that are designed to make sure the choices you make are functionally identical to the choices that are made for you without your knowledge. To it, justice is quite different from the task of distinguishing the guilty from the innocent; rather, it is a matter of separating those who are hard to prosecute (because they are rich or politically connected) from those who are easy to prosecute (because they are poor and disenfranchised). Arresting someone and then not convicting them—just because a person happens to be innocent—would be an inefficient use of police time.

It demands blind faith in progress

BASIC ARTICLES OF faith for all those who serve the technosphere include the following:

- Technology is inherently good.
- New technology is better than old technology.
- More technology is better than less technology.
- Anyone who stands in the way of more, newer technology is simply out-of-date.

The fact that each of these presuppositions is wrong about as often as it is right is of no consequence; these are not logical propositions but articles of faith, and to question them is to commit technological heresy.

Such blind faith precludes a reasoned approach to technology. With any activity, there is an optimum amount of it, and too much is just as bad as not enough. Drinking too little water leads to dehydration, which can be fatal; drinking too much causes hyponatremia, which can also be fatal. With any sort of progressive development, it eventually reaches the point of diminishing returns, where further effort in the same direction produces less positive results than before. Then it reaches the point of negative returns, where working harder at it produces less benefit. This principle applies equally well to economic development, national security, public safety, scientific research, educational standards... and, of course, to technology. But such obvious reasoning is anathema to the technosphere in its quest for infinite technological progress.

Look, for example, at the progression of electronic communications. It started with radio: in the evenings, the family would sit around the "wireless" and listen to news bulletins and entertainment programs. This already tended to displace other, perhaps more valuable activities, such as making and mending things, keeping oral history alive through storytelling, reading books (aloud to each other), playing games and discussing the day with the family. Allowing disembodied voices to invade the home introduced some amount of alienation into the family, because their focus, when together, was no longer exclusively on each other, but was now disrupted by the voices of phantom strangers.

Next came television, which added a dimension of hypnosis: now not only did family members listen to disembodied voices instead of to each other, but they no longer looked at each other either, because their eyes were glued to the ghostly images dancing across a flickering screen. The ability to broadcast moving images right into the home gave broadcasters sweeping new powers: now they could use images to belittle people and make them feel inadequate, to make them look down on each other and to entice them to pursue dreams of individualistic material affluence while disregarding the interests of the family and the community.

Next came the internet, initially in the form of the personal computer with a telephone dial-up connection. This was, at first, beneficial, giving people access to information they would otherwise never encounter. It opened up new freedoms for authors and other creative people, which were denied to them by the older, traditional models of publishing. But these benefits came with adverse effects: the family no longer participated in the same things because each person interacted with the internet separately and in different ways. Internet addiction started to be recognized as a serious condition in need of treatment, alongside substance abuse.

Next came mobile computing via tablets and smartphones. Now many people are virtually never without access to the internet, constantly sending and receiving messages. For many people, the little glowing screen has replaced the world around them. It is not the least bit unusual to see siblings gather at their parents' house during the holidays, only to sit in stony silence poking at their smartphones, each lost in his or her own world.

Their extreme level of internet addiction means that they will never venture into areas that lack network coverage and will voluntarily confine themselves within an open air prison whose walls are defined by the reach of wifi hot spots and cell phone towers. It also means that they will remain docile and subservient, and will do whatever is asked of them in order to avoid being cut off from the internet and from services such as Facebook and Twitter. For many of them, being cut off from social networking would qualify as a great personal tragedy resulting in severe psychological trauma.

This seems sufficient, doesn't it? Aren't we already far beyond the point where personal communications technology has reached diminishing, then negative, returns? But progress cannot stop, so let's keep going! After all, who could have predicted the smartphone's appearance and quick rise to dominance in 1981, when IBM unveiled its Personal Computer, which was priced at $1,565 ($4,290 in 2016 dollars) and featured 16KB of RAM, a 640 by 480 pixel 8-bit color display and two 5.25-inch floppy disk drives capable of storing 360KB each? Nobody but a crazed pie-in-the-sky techno-triumphalist!

And so let's give that techno-triumphalist a soapbox to stand on and listen to a short lecture about our seemingly inevitable techno-triumphalist future.

The next natural step after the tablets and smartphones is the wearable computer. Apple Watch and Google Glass are but tiny steps in that direction; a few more technical advances are needed to make the technology really useful and cheap. But when this happens we will find people wearing an integrated device that incorporates AR (augmented reality) glasses with accurate eye tracking, headphones, a microphone and a variety of other sensors such as an infrared camera and a laser range finder. It will, of course, include wifi, GPS and Bluetooth and run all of our favorite apps. After a short period of time, wearing these devices will become obligatory in a variety of settings, because without them we will be unable to see what everyone else is looking at. As a cost-saving measure, it will become more efficient to eliminate real world signage in favor of just posting virtual signs online as part of our augmented reality. Eventually this trend will spread to traffic signs, safety warnings and other required bits of public information. After this point, the AR glasses will become mandatory for all adults and most children when they are in public places, although few will ever take them off during their waking hours, finding being without them troubling and disorienting.

But wearable computers will prove to be just a clumsy transitional stage. The next advancement will include direct nerve tissue interfacing. The entire computing device will now fit inside a miniature capsule plugged into a socket cemented into one's molar tooth. It will communicate wirelessly with implants that will interface with the sensory and motor cortexes, the auditory nerve, the visual cortex and, of course, the pleasure center. Children will be implanted with these devices as soon as their milk teeth grow in. Although parents will be allowed to opt out of having this done, most will not, because the implants will be required by daycare centers. By then, the ability to entertain and control toddlers without the use of these implants will have been lost.

Next, it will be discovered that aside from the brain the human body no longer serves any definite purpose. Even most of the head is now of limited use, since functions of the eyes and the ears have been replaced by the implants. With sexual intercourse replaced by virtual reality porn and artificial insemination, the body will fall out of favor. Also, as people age, their bodies tend to develop medical problems that reduce their quality of life and shorten their lifespans, while their brains, at some basic level of functioning, can be kept alive quite a bit longer. And so it will become fashionable to have a certain procedure performed, called radical corporectomy, in which one's body is amputated below the neck, leaving just the head attached to an apparatus to keep the head healthy and comfortable.[1] This procedure will at first be performed as a life-saving expedient, but later people will choose to have it done as soon as they develop any half-serious medical problem. Medical insurance companies will be particularly quick to embrace radical corporectomy as a cost-effective alternative to many other operations, the surgery being simple enough—snip-snip, plug-plug—to be performed by a fully automated surgical robot.

In parallel, a movement will develop to virtualize people in their entirety, their heads included, by replacing them with computer simulations. At first this will be done to keep your loved ones alive once they have passed away, but later parents will decide that having virtual, simulated children is much less troublesome than having physical ones, what with all of the expense of giving them neural implants and later having their bodies amputated. People in their advanced years, fearing the onset of dementia, will opt to have their brains digitized ahead of time to avoid embarrassing themselves on social media.

---

1   This concept is from Alexander Belyaev's 1925 science-fiction book, *Professor Dowell's Head.*

And this will set in motion the final, inexorable trend in which actual, physical humans will be replaced with computer simulations of them. By then computing power will have progressed to the point where the simulations will bear an uncanny resemblance to the supposed original, being able to text things like "OMG!" and "LOL!" and exchange selfies of their simulated duckfaces just like the originals.

As a matter of efficiency, the simulated humans will only be made to function for the benefit of the few remaining non-simulated humans. And after the last remaining human is replaced by a simulation, it will finally become possible to turn off the whole thing. The technosphere wins; game over.

Its only alternative to infinite progress is the apocalypse

BLIND FAITH IN infinite progress is becoming a hard product to sell. Technology does improve, but does it necessarily improve people's lives? Are people more secure in their jobs? Are their jobs more fulfilling? Do their wages and salaries rise as technology improves? Are they healthier and happier?

These improvements may be true, to some extent, when it comes to managerial and technical professionals, but for the rest of us plummeting labor participation rates, stagnant wages and various other statistics tell a different story. Just one interesting fact—that economic conditions now force very close to half of all 25-year-old Americans to live with their parents—shows what this high-tech future holds for most people.

And so it becomes necessary for the technosphere to periodically apply some discipline in order to keep the dream of infinite technological progress in the service of humanity from starting to look a bit threadbare. The way this is done is by presenting any alternative to endless progress as an unmitigated disaster: it's either business as usual or the apocalypse. There are many different varieties of the apocalypse, featuring various combinations of asteroids, zombies, deadly viruses, space aliens, shark-bearing waterspouts over Los Angeles... the list is endless.

But there is one scenario—a realistic one—that you will hardly ever hear about. It's the one where people come together and decide that further investment in science and technology would be unhelpful and set up committees to de-technologize various activities and occupations to bring the amount of technology they use down to an optimal level. They may decide that libraries should, for the time being, keep electronic catalogs, with index cards available as a back-up, but that e-books should be done away with. Or they may decide that, to save time that's wasted in pointless communication, e-mail should only be delivered once a day, at 8 o'clock in the morning. They also set up some committees to thoroughly review and vet every single technological change that is proposed, to see what its unintended consequences could be, before permitting it to be implemented.

No, such a scenario would be considered entirely unacceptable, and an array of technology experts would be ready to make claims that it would "harm innovation" or "harm the economy." The idea that **the precautionary principle** is vitally important and should be followed at all times would be ridiculed because, you see, if there is an unintended consequence or two, we have the technology to fix everything... or, if we don't, then we'll surely develop such technology... once we realize that we have to... eventually. But questions of public safety or human needs must never be allowed to stand in the way of endless technological progress!

It always creates more problems than it can solve

THE FLAT REFUSAL of the technosphere, and of all who sail in her, to abide by **the precautionary principle**—that we should hold off on the introduction of a new product or process whose

ultimate effects and unintended consequences are either dis-
puted or unknown—means that it produces a constant overhang
of unintended consequences, big and small, to which technological remedies must be applied. In turn, these technological remedies
inevitably have unintended consequences of their own that are
often worse than the original problem. Moreover, for quite a few of
the unintended consequences there is simply no technology, either
in existence or in development, to remedy them.

Take, for instance, the problem of nuclear proliferation. Around
the end of World War II it became apparent that the industrial
capacity for destruction using conventional weapons had reached
such an extreme that new ways for preventing wars needed to be
sought. This realization resulted in the development of nuclear
weapons as the ultimate deterrent, preventing the outbreak of war
between the major powers through mutual assured destruction.
But the invention of increasingly powerful nuclear weapons—the
hydrogen bomb, the neutron bomb—along with more advanced
delivery systems, has continually degraded their deterrent capa-
bility because one side could wipe out the other in a first strike,
allowing no chance for retaliation, and the result has been an arms
race and an endless stockpiling of nuclear weapons by the USA and
the USSR.

One unintended consequence of the nuclear arms race was the
appearance of outlandishly huge stockpiles of plutonium in the
USA and the USSR. When saner heads prevailed and disarmament
talks resulted in an agreement to reduce the nuclear arsenals, a
solution had to be found to utilize the extra plutonium. The solution was to create mixed oxide (MOX) fuel for nuclear reactors,
which combined some plutonium with uranium (which, by the
way, was becoming more scarce, harder to mine and more expensive). Around the world a fair amount of our electricity is now
generated using MOX fuel that's laced with the world's most potent
bomb-making material.

This development resulted in several more unintended consequences. First, it made nuclear fuel much more expensive to

reprocess. Second, it increased the risk of proliferation of nuclear weapons. Third, it increased the risk of cancer from plutonium contamination in case of nuclear accidents (something that's already happened at Fukushima). And, to top it off, it further exacerbated the ever-growing problem of eventual long-term high-grade nuclear waste disposal, because the previously compact plutonium stockpile is now diffused among thousands of spent fuel rods. The interim solution to these problems is to simply keep the spent fuel rods at the nuclear reactor sites submerged in water. At Fukushima, the process of removing the fuel rods from the spent fuel pool, which had been damaged by explosions, was carried out at great expense and took several years.

But even the effort of moving the spent fuel rods to some place slightly safer is simply a way of "kicking the can down the road" and not even that far. There are no adequate long-term disposal plans for all of the high-grade nuclear waste anywhere, and very little is being done about this. Essentially, the entire nuclear industry blithely assumes that stable social and economic conditions will prevail over the many millennia which it will take for the nuclear material to decay. This is an untenable assumption and is but one example of the technosphere creating problems it is currently incapable of solving.

In the meantime, Fukushima is just one accident that has already happened, but there are sure to be hundreds of others: nuclear facilities that will be below sea level as glaciers continue to melt or deprived of cooling water as droughts caused by global warming reduce river flows. It is becoming apparent that our current technosphere isn't big enough or powerful enough to solve these problems. A much bigger technosphere could perhaps be able to solve all of these problems some day, but a much bigger technosphere would require a much bigger planet.

The nuclear waste disposal problem is a huge one, but an even bigger one has been caused by the burning of fossil fuels, which has increased atmospheric CO_2 concentrations from 280 parts per million to over 400. The last time there was so much of this greenhouse gas in the atmosphere was 3.6 million years ago. It has been shown that at that time the Arctic Ocean was largely ice-free and ringed by dense forests, and climate models suggest that even Antarctica may have been forested. Also at that time, ocean levels were 15 to 25 meters (50 to 80 feet) higher than today. Although politicians are still talking as if the rise in global average temperature can be limited to just two degrees Celsius, a more realistic best case scenario is 3.5 degrees—and that's if every country fulfills its pledges to cut emissions agreed to at the COP 21 meeting in Paris in 2015. And that would already be enough to trigger positive feedbacks such as Arctic methane release, which will drive the temperature even higher.

Leaving aside the question of whether such a world, in which tropical oceans are too hot to swim in, is conducive to the survival of any remnant of the technosphere or of humanity, what can the technosphere offer to mitigate the effects of this massive climate disruption? Well, it turns out that there is no shortage of people in the world with megalomaniacal, techno-narcissistic ambitions, and they weren't too shy to make some modest proposals. Their proposals go under the rubric of terraforming: transforming the entire planet.

One modest proposal that was tabled is to seed the oceans with trace elements that would spur gigantic blooms of plankton. The plankton would convert carbon dioxide to oxygen, returning ocean acidity back to normal and eventually reducing atmospheric carbon dioxide concentrations. Of course, there might be some unintended consequences. The bloom of plankton would also drive up the population of animals that eat plankton, and as they breathe oxygen and produce carbon dioxide they would rapidly de-oxygenate the oceans. Once they run out of oxygen and/or gobble up all the plankton they would die and create an anoxic environment, causing a bloom of anaerobic bacteria which would produce a layer of hydrogen sulfide under the ocean surface. Under certain conditions, inversions could cause hydrogen sulfide-laden water to well up, bubble up out of the water, drift onto the shore and kill every animal there. Isn't that a nice plot for a Hollywood horror movie?

Another modest proposal was to put into orbit a large number of giant mirrors which would shade the earth, bringing down global temperature. One immediate unintended consequence of this plan is that the industry that would manufacture and launch the space mirrors would burn lots of fossil fuels, adding more carbon dioxide to the atmosphere, accelerating global warming, and thus partially defeating its stated purpose of reducing global temperature. Another problem would occur if some number of space mirrors were to get hit by asteroids and start hitting other space mirrors, filling low Earth orbit with flying debris as per the Kessler syndrome (a.k.a. the domino effect). Even without any asteroid hits, some space mirrors would probably lose attitude control, start tumbling and irradiate the planet instead of reflecting sunlight away from it, periodically setting parts of it on fire. And then, of course, there is the problem of what to do once we run out of nonrenewable natural resources needed to manufacture and launch the space mirrors. Burn up?

Dr. Greg Laughlin of the NASA Ames Research Center in California proposed an even more ambitious plan: shift the Earth's orbit away from the sun by shooting asteroids at it...and barely missing each time. The transfer of gravitational energy from the asteroids to the Earth would gently nudge it up to a higher orbit around the Sun, where it's a bit cooler. "The technology is not at all far-fetched," Laughlin was quoted as saying. Do you look forward to having him and his colleagues shoot asteroids at you and just barely missing? The potential unintended consequences of this plan are perfect material for a Hollywood disaster movie. Remember, this is the same NASA that lost its Mars Climate Orbiter in 1999 because of failure to convert navigational data from Imperial to metric units while programming it. "Trust us," they will no doubt tell you, "we'll probably get it right this time." Try to imagine an asteroid accidentally parked in a decaying orbit around the Earth, threatening it for years until finally landing directly on Dr. Laughlin.

We could go on, compiling long lists of problems that are unintended consequences which the technosphere has caused and is currently, and will perhaps forever be, incapable of solving. But an even bigger problem is that there is now a large population that is abjectly dependent on the technosphere's services and will find it very difficult to survive without them. This dependence manifests itself in many negative ways:

- Over several generations, easy access to cesarean sections during birth have bred a population of women that, when deprived of them, are much more likely to die in childbirth, give birth to a stillborn infant or develop an obstetric fistula.
- Children who have grown up with vaccines and antibiotics have formed a population that would have very high infant mortality if deprived of access to them.
- Those who have grown up with computers, calculators, word processors, spell checkers, CAD systems and search engines would be at a loss if forced to produce a financial report or draft a legal or an engineering document using nothing more than pens and paper.
- Those who have grown up on purchased food would not know how to feed themselves if deprived of it, knowing neither how to grow food nor how get it through hunting, fishing or gathering.
- Those who have grown up with the idea that their security is to be left up to the experts would not know how to self-organize for their mutual self-defense if the police disappears or fails to act.
- Those who have grown up with unrealistic expectations of endless technological and economic progress, and whose natural instincts for autonomy, self-reliance and spontaneity have been bred out of them, will be largely incapacitated if thrust into a situation where they have to fend for themselves and look out for each other.

Luckily, these are problems that we don't need the technosphere in order to solve. Nature will solve them just fine. And the way nature will solve them is by letting those who can't adapt die and allowing those that can to survive.

Why it will fail

THE TECHNOSPHERE CANNOT go on doubling in size forever for a very simple reason: the planet is finite and the amount of nonrenewable natural resources it can provide is limited. It has already run up against some of these limits, while many others lie just ahead. Virtually all the sources of high-grade metal ores have been depleted. For example, gold is still being mined, but its concentration is now down to three parts gold per million parts of crushed rock. That rock takes a lot of energy to crush and sift, and gold mining now consumes an ever-increasing amount of energy per unit gold. And so it is with most other nonrenewable natural resources: the more diffuse they become, the more energy they take to produce. But the supply of energy is, in turn, finite.

Let's look briefly at the supply of crude oil, which is a key ingredient without which the technosphere cannot function. (Yes, a different technosphere—one that doesn't use diesel engines in locomotives, ships, trucks, electric generators, construction equipment and that doesn't burn some sort of petroleum-based fuel in most other forms of transportation—could theoretically exist. But it doesn't exist now, there is no time, no money and no spare energy to build it, and there are no sufficiently abundant and concentrated alternative sources of energy on which it could operate.) As early as 1970 the global supply of crude oil was predicted to peak in 2000. It peaked in 2006—half a decade later than initially predicted. Conventional crude oil—the kind that squirts out of conventional oil wells on dry land and can be produced cheaply—has been dwindling by around 5 percent per year ever since. As a result, at one point oil prices spiked at close to $150 per barrel and stayed high for several years, crushing economic growth in many parts of the world.

The technosphere managed to keep going by replacing this cheap-to-produce, conventional oil with expensive, exotic oil—deep offshore oil, shale oil produced by hydrofracturing, synthetic oil from tar sands and so on. To avoid running short of oil, the technosphere started channeling more resources into the energy sector by starving other sectors. But this only worked for a few years, and until oil dropped to around $30 per barrel—which is less than it costs most of the unconventional producers to produce a barrel of oil. As a result, many oil producers will fail through bankruptcy, the oil supply will fall, and the oil price will spike again, delivering another blow to the economy.

This really isn't that complicated—but there are plenty of people who refuse to understand it. The phenomenon of peaking crude oil production is called Peak Oil for short; it has been carefully tracked for a long time, and it is quite real. But squandering lots of capital on unconventional oil (which, as it turns out, is too expensive to allow the technosphere to continue to expand) was enough for some people to declare that Peak Oil is dead. And now that oil supply has temporarily exceeded oil demand by a small margin causing prices to crash, they have declared it doubly dead.

This is reminiscent of other such declarations. An extreme cold spell is enough for some people to start declaring that global warming is a hoax. The subtlety of the situation—that a rapidly warming Arctic is causing the temperature gradient between north and south to decrease, in turn causing the jet stream to meander, driving cold Arctic and Siberian air far south—is lost on them. But you should not be fooled, not by a temporarily low oil price nor by a temporarily low outside temperature. Yes, there is quite a lot of public relations money behind such declarations, but that's all there is behind them.

"They will come up with something!"

THIS HAS BECOME something of a mantra for those trapped within the technosphere as soon as they become aware of the fact that the technosphere's continued survival and, by extension, theirs, is by no means guaranteed. "But we have the technology!" goes one impassioned outcry. "We can innovate our way out of anything!" goes another.

And then there are some sobering bits of reality. One of them is the very high correlation between economic growth, population growth and growth in the consumption of crude oil. This is not based on some model or a theoretical prediction; this is an observation. You may argue that correlation is not causation. You can also believe that a car going faster when someone steps on the accelerator is purely a statistical artifact unrelated to the underlying mechanism, and you will be proven right every time the car runs out of gas or its front bumper encounters a utility pole. So too here: every time energy got scarce (as with the Arab oil embargo of 1973) or too cheap (as in 2015–16) or too expensive (as in 1980 and 2009) this resulted in major economic disruption. But I would still argue that the car primarily moves because of gasoline (unless it's downhill, but then it doesn't move very far) and that the technosphere primarily exists and grows because of crude oil.

Those who champion the theory that "they will come up with something" need to realize that "they" have been coming up with something for a long time now, and that this something amounts to pretty much the same thing every time. Energy alternatives have been explored practically forever, and the breakthroughs are rather insignificant. Photovoltaic cells became cheaper and more efficient; large wind turbines became more advanced and proliferated along windy coasts and ridges. Other ideas, such as thorium-fueled nuclear reactors and nuclear fusion, are, and most likely will forever remain, technologies of the future.

One fact that should somewhat temper their enthusiasm is that while an extended period of record-high oil prices has succeeded in pushing up the share of renewable energy sources (hydro, wind and solar) to just over 20 percent of global electricity generation, the rest of it still comes from coal, natural gas, nuclear and oil. And now that oil is cheap again, the interest in investing in expensive alternative energy sources has dwindled. Who would want to invest in expensive energy when energy is so cheap?

Another important limitation on the role of renewables is that only a tiny fraction of transportation fuel, which accounts for close to half of all energy consumption, comes from renewables; the rest comes from oil. Virtually all of the transportation infrastructure, along with a lot of other industrial machinery, runs on diesel oil, jet fuel or bunker fuel, made from crude oil, and there is absolutely no chance that all of this capital equipment is going to be replaced with something that runs on renewable energy any time soon.

A "transition to sustainability"—whatever that means—would, it is generally conceded, require major scientific breakthroughs. But who would make such breakthroughs? Scientists are reasonably well provided for in their work, but only if it raises the technosphere's prestige. If they stick to the sort of boring applied science that may or may not someday result in some useful bits of technology, they are unlikely to find as much support. And if they attempt to conduct basic research—the sort of science that explores the world for no conceivable practical purpose whatsoever—they are very much marginalized.

Science started out as a sincere expression of curiosity—of marveling at the world and trying to comprehend it—without any practical applications in mind whatsoever. In fact, the idea that learning should serve some function in the marketplace was abhorrent to the ancient Greeks, as well as to ancient scholars in many other countries. But now science is largely the task of looking for the car keys under the streetlamp where you might be able to see them, the streetlamp being high-status, cutting-edge technology and the car keys a large corporate or government grant.

Wherever we see high-profile, high-budget science, we see a showcase of new technology, with the science used for propaganda purposes to glorify the new technology. It was the availability of powerful superconducting magnets and data acquisition systems to record the results of relativistic particle collisions, together with the supercomputers to model them, that made it necessary to look for the Higgs Boson, which was given the humble nickname "the God Particle." It was the availability of new gene sequencing technology that necessitated the study of the human genome. It was the development of advanced robotics that made necessary the launching of unmanned space probes to crawl about on Mars and poke at rocks that are of no practical concern to anyone. Scientific curiosity now comes down to being curious about just how much additional prestige one can get from using some fancy new gadget. I spent six years working in high energy physics, and I can tell you that scientific research is not driven by curiosity; it is driven by ego. The money is not that good, the amount of effort it takes to get recognized is ridiculous, and so the motivating force behind it is the outlandishly huge, incredibly fragile scientific ego.

Most high-profile, high-budget science has to do with dead things, with living things relegated to second-class status. Physics and chemistry rule; biology is only high-profile if it can be reduced to biochemistry. For example, cancer is a phenomenon of living cells, but cancer research treats it as a predominantly molecular phenomenon and searches for molecules that can kill cancer cells without also killing the whole patient. Another example: reproduction is a cellular process, but every effort is made to reduce it to looking at DNA, which is a molecule, and essentially treating it as data. It seems that almost everywhere dead things take precedence over living things, and if something isn't reducible to physics and chemistry, can't be made to function as a machine and can't be analyzed using mathematics, then it is considered an inferior subject for scientific inquiry.

This tendency shows up in the funniest places. For example, in the US, linguistics—the study of human language and associated cognition—took a truly bizarre detour thanks to the efforts of Prof. Noam Chomsky of MIT. Instead of looking at what's actually interesting about language, which is semantics (the study of meaning) and how it relates to epistemology (the theory of knowledge) and phenomenology (how we perceive reality), Chomsky seized onto a particularly boring aspect of language—syntax (the study of word order in sentences)—because it could be analyzed using formal, mathematical methods. Only really knowing English, Chomsky used English as the basis of his research, neglecting the broader study of both language and languages. As a result, several generations of American linguists grew up learning almost no linguistics, but instead engaged in formal syntactic analysis, struggling to derive the principles of a "universal grammar" that was hypothesized to be encoded into the human DNA. But human language cannot be reduced to mere formalism, and so after doing quite a job on American linguistics Chomsky gave up and went into politics instead. It was a safe choice: since American politics had been well and truly trashed before he got to it, his participation in it has proven to be entirely inconsequential.

And so we will have all the fruits of physics and chemistry—nuclear fusion experiments, space probes, nano-this and genetically engineered that—but wild, self-organizing living things will get short shrift, and when it comes to "sustainable technology" (whatever it may turn out to be), well, we won't have much of that, and even if we did, we wouldn't use it because it won't be sexy enough. Look at the bits of it that we do have—humble adaptations such as rooftop water collection systems and solar water heaters, composting toilets, architectural techniques that reduce heating and air conditioning needs and so on. They are hardly ever used and are sometimes even made illegal!

The few little sops to sustainability that do make it through—bike lanes, a few solar panels and wind generators, recycling—are mostly just bits of propaganda: "Look, we did come up with something!" Other efforts—such as reusable shoddy polypropylene shopping bags on offer at supermarket checkouts, into which you place little containers made of low-density polyethylene, vinyl, polyvinyl-chloride and other plastics with bits of industrially produced food inside them—are even more pathetic.

And so the answer is that, indeed, "they will come up with something"—nuclear fusion, thorium reactors, super-efficient batteries, synthetic oil made from algae, tidal energy, geothermal energy, space mirrors beaming energy down to Earth in the form of microwaves, energy from helium-3 mined by robots on the Moon... They will come up with energy from oceanic clathrates, which is methane ice found at great ocean depths. And let's not forget abiotic oil—based on the cock-eyed theory that crude oil isn't made out of dead organic matter over many millennia but is secreted by the Earth directly. They have come up with all of these already, and they will come up with all of them again, over and over, forever prompting people to exclaim: "Ah, look, they came up with something!"

But must we fail with it?

WE ARE NOW witnessing the early stages of what happens when the technosphere gets hungry. Commodity prices spike, then crash, in increasingly rapid alternation. Labor participation rates (which are far more indicative of social reality than the doctored-up rates of unemployment) continue to fall.[2] Well-paid, full-time jobs in production are being replaced with low-paying part-time service jobs with no benefits. Retirements are cancelled; promising new careers fail to start.

As the economic core attempts to sustain itself, the economic periphery—the developing and declining economies, and the outlying regions of developed ones—is being starved of resources. Mountains of bad debt, which will never be repaid under any conceivable scenario, are being piled up by governments, corporations and individuals, in what will turn out to be, in the end, a futile effort to paper over an ever-widening chasm between what has been promised and what can be delivered. This effort is gradually converting what were once safe and productive investments into worthless pyramid schemes.

---

2   According to the US Bureau of Labor Statistics, as of 2016 there are 159,286,000 people in the labor force, of whom 7,966,000 are unemployed. There are also 93,900,000 working-age people considered "not in labor force." If we define "unemployment rate" as "percentage of working-age people who are not working," then the US has 39 percent unemployment.

In the meantime, the biosphere—on which all of us, technosphere and all, depend for our continued survival, is in increasingly
poor shape. Large stretches of the ocean now contain more floating
bits of plastic than plankton. Marine organisms mistake the plastic for food, swallow it and starve. This may not sound dire—why
not have plenty of both plankton and plastic?—but the plankton is
dwindling, and it happens to make a lot of the oxygen we breathe.

The warming of the atmosphere—initially caused by the burning of fossil fuels, but now being exacerbated by positive feedback
such as Arctic methane release and widespread forest fires—is disrupting the weather, causing increasingly destructive storms and
endangering our ability to grow crops. The rising ocean levels and
the more powerful storms are eroding coastlines, making a great
deal of valuable oceanfront property uninsurable and ultimately
worthless. Thus, even if the technosphere were to somehow magi-
cally disappear (causing, by the way, a staggering loss of life) what's
left of the biosphere may turn out to be too severely disrupted to
sustain those of us who survive.

But we are an ultimately adaptable species, and I see no reason
to give up all hope just yet. My hope is that the technosphere will
fail and that we (at least some of us) will survive its failure. Further-
more, it is my hope that the technosphere won't fail all at once, but
in stages, giving us a chance to scavenge from it what we will need
to go on.

What would its success look like?

IF WE ASSEMBLE this entire picture of the technosphere, what
we see is an emergent global intelligence that hates all forms
of life, likes physics and chemistry, hates anything that it cannot dominate or control, is adept at using humans for its own
purposes, but is quite ready to kill them when they are no longer needed or when they get in the way, which it can easily do
because its most advanced and effective technologies are its killing

technologies—conventional, nuclear and chemical weapons; germ warfare; and political technologies that send people into battle.

On Earth, this entity has consumed most of the things it likes—easy to get at fossil fuels, high-grade ores and other concentrated mineral resources, fresh water—and is setting its sights on other planets. While it is unlikely that humans or most of the Earth's other higher life-forms could live on any planet except the one on which they have evolved, or could even survive the long trip to get to any other planet (because of cosmic radiation), their presence will not be necessary. Indeed, most recent efforts at space exploration have involved remote sensing techniques and automation—with an outright obsession with finding technically exploitable worlds inside and outside the solar system—and robotic exploration of this solar system using unmanned probes.

Manned space flight is still being talked about now and again, but the only ones who still have the ability to venture into space are the Russians.[3] In turn, Russian cosmonauts have only ever made it as far as low Earth orbit. The Americans claim to have made it to the Moon, but we will have to wait and see whether or not that was actually a clever hoax filmed by Stanley Kubrik, as many have suggested. Whether it actually happened or not, it was little more than a publicity stunt. The fact is, humans are rather useless and burdensome in space, while robots do quite well.

Unfortunately, from the technosphere's point of view, humans are still a necessary part of the robot manufacturing process. Nothing can get off the ground, so to speak, without some amount of help from human engineers and technicians. The natural outcome of the recognition of this fact is the great push to promote robotic manufacturing, 3D-printing technologies, nanotechnologies, biomimicry (making machines emulate living organisms) and all other forms of automation. The ideal, then, is that the technosphere would create numerous spores in the form of space probes

---

3   The Chinese manned space program is also showing promise.

and launch them toward all the technically "habitable" planets it finds in our rather tiny immediate neighborhood of the Milky Way galaxy. Once these spores land, they will 3D-print all sorts of mining and manufacturing technologies with which to build— what else?—more space probes, of course, without any human interference.

But what would happen if one of these probes landed on a planet that has a biosphere—is contaminated with living organisms, that is? Well, the probe would probably carry along some biocidal technologies, just in case, so that it could quickly wipe them out.

I certainly hope that nothing ever comes of this idiotic megalomaniacal techno-fantasy. I prefer to imagine that no probe ever gets constructed or launched, or that all the probes either fail to reach their destinations, or get damaged by collisions with tiny asteroids and fail to land, or do land but fail to achieve anything once they do. Except in one case: there should be that one special planet, where a probe does land, but then the local insects find the biocide inside it delicious. And then the sentient beings that inhabit that planet (sentient only when sober) find it and go on to fashion it into a perfectly serviceable moonshine still.[4] That, to me, seems like the best possible outcome for the technosphere in this the best of all possible universes—shrunk to size and put to good use.

The Anti-Gaia Hypothesis

IT CAN SOMETIMES seem that the technosphere thwarts its own purpose. What sense is there in wasting resources on weapons, when there is already enough war materiel to kill all of us several times over? What sense is there in contaminating the environment

---

4   This idea came from a brilliant but now long-forgotten winning entry in a Soviet science-fiction short story contest published in a Soviet youth magazine in the early 1970s.

with long-lived chemical toxins and radionuclides, producing high rates of cancer in the technosphere's human servants? What purpose is there in fostering extreme levels of corruption in government and in banking, or in creating conditions for extreme social inequality? How does it help the technosphere grow stronger and more controlling to provoke international conflicts and split up the world into warring sides? Are these all failings, or are they just little problems that are too small to matter? Or—here's a shocking thought!—maybe they are all perfectly on strategy as far as the technosphere is concerned.

If we look closely, we will discover that all of these manifestations of the technosphere, although on a superficial level appearing to be problems, are, in fact, helpful to the technosphere in many interrelated ways. They help the technosphere grow, become more complex and more fully dominate the biosphere. This is true in far too many ways to trace out all of them, but let's just consider a few of the more important "problems"—the ones I mentioned above.

With regard to cancer, it would seem that minimizing the rates of cancer by keeping carcinogenic chemicals and radioactive contamination out of the environment and by eliminating microwave and ionizing radiation would be a very good idea. However, this turns out to be suboptimal from the technosphere's point of view. First, this would violate one of its prime directives by prioritizing the interests of the biosphere above its own technical concerns. Second, this would limit the need for technical intervention. Cancer treatment is a tour de force for the technosphere, allowing it to use its favorite techniques— chemistry (in the form of chemotheraphy) and physics (in the form of radiation therapy)—to kill living things (cancer cells, that is). Third, it would forgo the opportunity to exercise control over people and to force them to serve and obey it, lest they find themselves deprived of very expensive, supposedly life-saving cancer therapies. What is optimal for the technosphere, then, is a situation where everybody gets treatable forms of cancer and where nobody believes they have any hope of surviving without chemo and radiation therapy. The technosphere likes us to be patient with it, and medical patients are patient by definition.

When it comes to fostering extreme levels of corruption in government and banking, this again seems at first counterproductive: wouldn't a lawful, efficient financial sector and a transparent, moral government be expected to produce better results? Yes, but results for whom? Moral governance and proper banking regulation would serve the purposes of... humans! That's right, it would be bits of the biosphere reaping the benefits again! And so it is far more efficient, from the technosphere's perspective, for the major banks to corrupt government officials by funneling money to them through a variety of schemes and to have these officials then refuse to regulate those same banks or to prosecute them for their crimes. Once all of this corruption becomes endemic and entrenched, the allegiance of public officials is no longer to the tricky, willful living entities known as "constituents" but to abstract tokens of wealth, which are much easier for the technosphere to manipulate to its fullest advantage.

Finally, wouldn't world peace and a benevolent and unified world government be of much more use to the technosphere than having humanity continually split up into warring sides? Perhaps, but what would that do for enhancing the technosphere's ability to murder people? When the great nations have to constantly prepare for war, they are forced to arm themselves, and in order to do that they have to industrialize—to develop and maintain an independent industrial base. Were it not for the need to keep up with the arms race, some nations might prefer to forgo industrialization and remain pastoral or agrarian, but because of the threat of war, the choice is between industrialization and defeat.

War has other benefits as well. War requires swords which, once war is over, are beaten into ploughshares, which leads to increases in agricultural efficiency, which in turn make peasant labor redundant and drive peasants off the land and into the cities, where they are forced to work in factories, thus driving more industrialization. War offers an easy way for industrialized armies to exterminate or enslave nonindustrial tribes, who would otherwise be setting a bad example by living happily outside the technosphere. Lastly, without a powerful war machine people would be able to self-organize and provide for their own security, making them harder to control. The existence of powerful military weapons makes it necessary to put security in the hands of tightly controlled, strictly disciplined, technocratic, hierarchical organizations—just the sort the technosphere prefers.

Thus it appears that the technosphere, viewed as an organism, possesses a sort of emergent intelligence: it knows what it wants, and it can figure out nontrivial ways to get it. If this claim seems like an outlandish conjecture, then compare it to James Lovelock's **Gaia hypothesis**. According to Lovelock, all of the living organisms that inhabit the Earth's biosphere can be viewed as a single super-organism—a complex, self-regulating system that interacts with the inorganic elements of the planet in such a way as to make it habitable. Its basic functions include regulation of temperature, ocean salinity and the atmospheric concentrations of various gases. This ability of the biosphere to maintain homeostatic equilibrium and to restore it in case of disruptions in the form of, say, volcanic eruptions and major asteroid impacts can be viewed as evidence of an evolved intelligence that strives for the greatest possible complexity and diversity of the web of life. Although somewhat controversial and not directly testable, the Gaia hypothesis is taken quite seriously in a number of academic disciplines.

Taken in this context, my hypothesis—let's call it the **Anti-Gaia hypothesis**—seems rather less outlandish. It is that the technosphere, having risen up on top of and in opposition to Gaia and the biosphere, possesses a certain primitive emergent intelligence that has allowed it to grow in complexity and power and to dominate the biosphere to an ever-greater extent.

Unlike Gaia, which is an organism unto itself, the technosphere is a parasite upon the biosphere, using living organisms as if they were machines and striving to replace them with machines as much as possible. This is perfectly obvious in industrial agriculture, which replaces complex ecosystems with the machine-like simplicity of chemically fertilized monoculture. The factory farm, in which animals are confined in a sort of mechanized hell, is a perfect example of how the technosphere prefers to treat higher life forms. When it comes to us humans, the best example of technosphere's influence is the modern corporation, in which people are incentivized (and in fact required by law) to act as perfect psychopaths, blindly pursuing shareholder profits to the neglect of all humane concerns. In politics, the technosphere gives rise to political machines, which treat voters as if they are laboratory animals, conditioning them to press certain voting machine levers in response to well-calibrated mass media stimuli.

Also, unlike Gaia, which strives to maintain homeostatic equilibrium, this intelligence strives for *disequilibrium*—for continuous growth, which, on a finite planet with limited stores of nonrenewable natural resources, is an obvious dead end—"dead" as in "extinct." To compensate, the technosphere dreams (with the help of certain humans who are in thrall to it) of universal conquest: it dreams of breeding a race of self-reproducing, space-faring robots. It dreams of leaving this exhausted, devastated planet behind and of colonizing other worlds—ones with lots more nonrenewable natural resources for it to mindlessly squander and, crucially, whole new biospheres for it to dominate and destroy. This last bit is very important because the technosphere's existence loses all meaning without living things it can force to act like machines. Without a biosphere to subjugate and destroy, the technosphere becomes just a blind, deaf robot whistling to itself in the dark. Without the miraculous, wondrous goodness that is life, the technosphere cannot even aspire to being evil—only banal. Widgets in space! Yawn...

2

WHAT IS AT STAKE?

Just how bad is it likely to get?

ALL OF US depend on the biosphere for our survival: for the air we breathe, the water we drink, the soil in which we grow food and the complex, ever-evolving community of microorganisms that inhabit our bodies and account for several pounds of our body weight. Disruptions to the biosphere pose multiple mortal dangers to us: global warming spreads tropical diseases further and further north; rising ocean levels are predicted to drown the coastal cities where nearly half of us live; disappearing mountain glaciers are turning arable land into deserts, threatening starvation.

Most of us also depend on the technosphere for our survival. If the lights went out, the public water supply failed, transportation fuels became unavailable and so on, most of us would lose access to food, medicine, heat and air conditioning, and would starve, suffer from dehydration and either hypothermia or heatstroke, fall sick and die. Without communications and transportation networks we would find ourselves stranded and unable to communicate.

It is also by now quite clear that the technosphere dominates, disrupts and kills off the biosphere in a myriad interrelated ways— too many to enumerate. To mention a shocking few:

- The oceans are becoming quite literally trashed. There is the plastics plague which has inundated the oceans with tiny, long-lived bits of material which, as they decay, release toxins into the marine environment. There are the increasing ocean acidity and water temperatures, which are imperiling shellfish and coral. Now add all the chemical toxins, such as the 1.84 million gallons of Corexit oil dispersant that BP used after the Deepwater Horizon disaster and the fertilizer runoff from farms and lawns, which has caused anoxic dead zones to appear and spread. All of this is forcing the oceans to revert to a primordial state dominated by bacteria and jellyfish. How will coastal and island populations survive if deprived of the sea as a source of food?

- There is the nuclear contamination issue: all of the long-lived radioactive materials from both nuclear weapons production and nuclear power will remain dangerous for far longer than the maximum lifetime of any conceivable civilization, longer even than the maximum expected lifetime of the human species. As facilities that house nuclear material are abandoned and fall into disrepair, plumes of radioactive contamination will spread across various areas of the planet. How could post-industrial societies be expected to be able to track and map the nuclear contamination as it is gradually spread by winds and currents, migrating bird species, smoke from forest fires, storm runoff and so on?

- The level of climate disruption due to the burning of fossil fuels may have already resulted in unstoppable positive feedback that will put the Earth's climate in a state that will nullify all our efforts at agriculture. Humans developed agriculture roughly 10,000 years ago and started the chain of events in which civilizations rose and fell, culminating in the present global industrial civilization. According to the climate record deduced from ice cores, fossilized tree rings and other sources of evidence, these last 10,000 years were also a period of unusual climatic stability. This was not a coincidence: without this stability there would have been too many failed harvests to maintain a stock of seed grain, and agriculture in its most common form—the tilled monoculture of annual plants—would not have been possible. And now that we have entered a geological epoch that some are calling the Anthropocene—because it is so significantly impacted by human activity—agriculture is once again likely to become a losing proposition. How are human populations going to survive if the usual methods for growing staple foods can no longer be relied on?

- Infectious disease control has prevented many deaths and resulted in a very large population, but antibiotics have been overused by doctors and livestock farmers alike. Now bacteria are evolving antibiotic resistance faster than new antibiotics can be invented, tested and made available. Some medical experts predict that antibiotics will become useless in as little as a decade, leaving a large population of both humans and domesticated animals that, through the use of antibiotics, has inadvertently been bred to be defenseless against infectious disease. How will human communities and families be able to cope with the large, sudden increase in morbidity and mortality that will occur when industrial medicine fails and disease loads rise?

- Last, but by no means least, the advance of technology has produced a human population that is far more helpless and dependent than any human population before, one that is unable to survive when exposed to the elements, or travel long distances on foot, make its own tools, construct its own shelter, clothe and feed itself without outside assistance, treat diseases with substances available from the environment, or teach its children to survive on their own... How will these people, who have been conditioned since birth to expect to be taken care of by a vast industrial machine, respond to suddenly being forced to rely on their own wits and physical strength to survive? How many of them will not even try and simply await a rescue that will never come?

It follows that if the biosphere wins the struggle and the technosphere fails and disappears, *many* of us will die, but if the technosphere wins and kills off what's left of the biosphere, then all of us will die. That is the difference: destroying the technosphere is a suicidal move for *most* of us; letting it go on is a suicidal move for all of us.

Does it have to be this way? I certainly hope not! But what is the choice? Do we really have to choose between genocide and extinction, or is there the possibility of a third choice? I want to believe that there is. The task, as I see it, is not to destroy the technosphere, nor to allow it to grow uncontrollably and then, just as uncontrollably, fail. The task is to *shrink* it down to a few well-chosen essentials. This means depriving ourselves of many of our habits, luxuries and comforts. But that's just what's on one pan of the scale. What's on the other?

The bad effects of the technosphere are by no means limited to the environment: its effects on us are just as bad, if not worse. As we shrink it, we will gain everything that it has taken away from us: autonomy in decision-making; unstructured, unscheduled time; a relatively stress-free existence; the ability to live close to nature, to spend time with those we care about rather than with strangers and to make the things we need instead of shopping for things we don't … and, last but not least, to have hope for our future.

Remembering who we are

LET US PUT ourselves into the broader context of our history as a species. The current industrial civilization is a mere blip in our long history, which started with *Homo habilis*, the first tool-maker, some 2.8 million years ago. On this time scale, the entire episode of civilization, during which we developed agriculture and cities, accounts for 0.3 percent of it, while the two centuries since the start of industrialization make up a vanishingly small 0.01 percent. If we look at the entire existence of humans as a single day, then we invented agriculture around 10 p.m., and industrialized around 11:57 p.m. If we take the present moment as the dawn of the new day, then it's quite likely that the industrial episode will be over in less than a minute.

It is useful, therefore, to recall who we really are, the last three minutes of our history notwithstanding. Let us abandon the science-fiction idea that the last few seconds of these three minutes are our lift-off phase and that we are off to colonize other galaxies; we are not. So far all we have managed to do is blast a few widgets into the cold darkness of space, and we might launch a few more of those, but then that is going to be the end of our "space-faring" and "star-steading." Let us instead consider a future in which the current industrial blip looks like a momentary bout of planet-wide insanity, swiftly terminated by nonrenewable natural resource depletion and environmental devastation. Let us concede that it will most likely be followed by an equally quick reversion to norm—a well-equipped, intelligent, enlightened, long-lasting one if our efforts at shrinking the technosphere succeed, or a chaotic, calamitous and short-lived one should we fail.

What should we consider normal?

WHAT COULD WE say about ourselves that would be close to universally true, outside of the current industrialized context? Below I will list some generic properties of humans which should be uncontroversial but of course will be, because for many of us our perceptions of what is normal have been warped by life within the technosphere. The values inculcated in us are the ones the technosphere chose for us, to foster dependency and to make us easier to control. It wants us to be atomized, lone individuals because individuals cannot stand up for themselves nearly as well as tight-knit groups. It wants us to be dependent on it in as many ways as possible because dependent people are subservient people. It wants to take away our decision-making abilities, our judgment and our discretion and to put them in the hands of experts or, better yet, robots and algorithms running on internet servers.

Let us, then, start with what should be the least controversial: all humans are very closely related. There are now no hominid subspecies; we are all just *Homo sapiens*. Biologically speaking, we are all pretty much cousins. To ascribe a significant genetic component to constructs such as race and ethnicity is to ignore a mountain of evidence that these constructs are social concoctions with no basis in biological reality. But something akin to breeds does exist if natural selection and natural variability are allowed to exert themselves. Put humans in a hot, sunny place, and some 10,000 years later they will have dark skin; put them some place where it's too cold year-round to go naked, and some 10,000 years later they will have again lost their skin pigmentation. Make them chase down game in the savanna, and they grow tall and lanky; make them row kayaks amid ice floes and sit out polar winters in igloos, and they become thickset and squat. But if they all breed together, then rather quickly you get back to a medium-beige, medium-height typical human, just as, if you allow dogs to breed however they wish, they quickly revert to the typical pointy-nosed, curly-tailed medium-sized "yellow dog."

The second least controversial observation worth making is that it is normal for humans to develop a tremendous diversity of cultures. Each unique type of natural environment requires its own set of unique cultural adaptations, but, beyond that, every little band and tribe tries to be different from all the others just because it wants to—because being different reinforces group identity and loyalty and makes it difficult to switch groups. There is generally a very low tolerance for strangers, and human bands and tribes tend to deal with outsiders as groups, not as individuals. Individuality is generally only allowed to express itself within the group; outside of it, what is important is the ability to present a unified front.

Thus, it can be said that humans are naturally separatist and try to spread out across the landscape to avoid members of other tribes. But since bands and tribes need to interbreed in order to avoid inbreeding, there are some common techniques for selectively breaching intertribal barriers. One is bride-snatching: the oldest form of marriage is "marriage by abduction," and it still persists in a surprisingly large number of cultures, although it has mostly devolved into "mock abduction." This ritual requires the bride to protest, but not too loudly, or the bridegroom and his friends run the risk of a non-mock beating. There are a couple of other methods as well. One is to exchange children. This allows the two children to grow up as bilinguals, who are valuable if the bands or tribes ever need to trade, form an alliance or otherwise work together. Another is by taking captives. Slaves aren't all that useful outside of industry or agriculture, but they can be used as breeding stock or as translators.

Next, with some notable exceptions, humans tend to have crisply defined gender roles. Children are often allowed to do whatever they like, but the boys usually emulate their fathers, and girls their mothers. This makes for efficient parenting, because the amount of practical knowledge children must informally absorb from their elders is too large for everyone to be taught everything. There is usually a rite of passage that separates childhood from adulthood, and after this rite of passage the gender roles tend to become rather strictly defined. The old cliché that men hunt while women gather is absolute nonsense because both do both (generally true of trapping and gathering, while less so of hunting). Nevertheless, gender roles tend to be distinct. Notably, both genders exercise leadership: the men overtly through authoritarian actions and commands; the women covertly through persuasion, conspiracy, passive resistance and guile. But the actual locus of power is usually the family hearth, ruled by a woman, and few men are stupid enough to issue orders they know will be resisted.

There are other fairly uncontroversial near-universals as well. Humans tend to be monogamous: like otters, beavers, turtle doves, gibbons, swans, wolves, bald eagles, prairie voles and barn owls, they breed for life. They tend to be intensely private about their sex lives and start insisting on some modicum of privacy from a young age. They tend to develop a few close friendships outside of their families that persist over their entire lives.

Somewhat more controversially, humans tend to be territorial and, even if they are nomadic or migratory and wander over a large territory, their sense of self is deeply rooted within the natural landscape. Certain of its features are often considered sacred—a particular rock, a grove or a spring. They regulate their interactions with nature and with each other using a set of taboos and unwritten rules. And they maintain an oral history, a cosmography and a mythology, which are passed down from generation to generation as epic poems, songs and stories, some of which persist for thousands of years.

More controversially yet, like plenty of other animals, humans kill their own kind—for all kinds of reasons. Some are even cannibalistic. Warfare offers a straightforward, natural way to decrease population pressure on the environment because crowding instinctively increases many animals' propensity for violence, humans included. Warfare can be used for all sorts of purposes—defending territory, enforcing a relationship based on tribute, even wholesale genocide of groups whose customs are considered disagreeable. When two tribes fight over territory, it is not unusual for the winners to dispatch all the males on the losing side and to take the females for themselves. Raiding one's neighbors is also an old favorite, and back-and-forth raiding is sometimes used to alleviate various kinds of accidental inequality. Although fratricide is generally taboo and so is patricide, it is not entirely uncommon to passively let old people starve in times of famine.

Lastly, and perhaps most controversially so far, humans generally have a very low tolerance for abnormality, and it is an unfortunate but indisputable fact that compassionate treatment of those who are viewed as abnormal is very far from a human cultural universal. Infanticide is a common way of getting rid of infants with birth defects. Physical perfection is usually very highly prized, and deviations from it are treated quite harshly. The weak and the infirm, those with chronic ailments, aberrant personality traits or perverse sexual tendencies all tend to be treated quite differently from the rest—not as full members. Unless they have great special talents, they are not valued, and they can easily be abandoned or neglected and, in the harsher societies, banished or even killed. Those who are considered "freaks" are often mocked and abused. Far from being arbitrarily uncharitable, such attitudes towards the abnormal and the handicapped are of practical survival value. The act of survival is so arduous and demanding, both physically and mentally, that the typical human band or tribe must resemble, for lack of a better metaphor, a sports team, and nature does not organize any sort of Special Olympics for us.

And now comes the most controversial observation of all. All of our vaunted civilizational values—including human rights, representative democracy, the rights of minorities be they racial, ethnic or sexual, the rights of the handicapped—have no place in nature. They are part of a culture—one single very special culture that has had an exorbitant amount of influence over the entire planet because it is optimal for the technosphere. No matter how much we treasure liberalism, humanism, gender equality, human rights, democratic principles, minority rights, rights of the handicapped, "responsibility to protect"[1]—no matter how much in love we are with all of that, when the technosphere fails, so will this culture.

Some of us have grown up thinking, along with Thomas Jefferson, that there are certain "truths" that are "self-evident," which include the "unalienable right" to "life, liberty and the pursuit of happiness," and we need to take a step back and reflect. To start with, nothing is "self-evident." That's just a pompous but empty phrase, because an established fact can be used as evidence in support of *another fact that is yet to be established*, but it cannot be evidence of itself—that's called a tautology, and it doesn't advance

---

[1]   "Responsibility to protect" (R2P) is a dubious political principle according to which when Tribe A is busy slaughtering Tribe B to a man, Tribe Q from across the world has the right to intervene to save Tribe B from extinction, in spite of Tribe Q having no standing in the matter of said slaughter.

an argument. As for the rest of it, see for yourself: do you see any examples where any of these rights seem very much "alienable"? Are there any murdered or imprisoned or miserable people in the world? Well, what about their rights, then? Don't you have a "responsibility to protect" them? Leap into action forthwith, then, and right these wrongs!

Perhaps before leaping into action you should look around first, to see what you might encounter. Do you see any men who claim to have an unalienable right to cut out a man's heart, eat it in front of a video camera and post the result on the internet? And do you see your national leaders doing much of anything to stop them? Perhaps the best you can do is not be part of the same tribe as these men and not let them anywhere near your own tribe. And before you can even do that, you will need to figure out who is in your tribe and who isn't.

In the end, all these vaunted principles and values, which so often go under the label of "Western," will turn out to be the shibboleths of a culture that is tethered to the technosphere and will die with it. If you like them, feel free to keep them, but be warned that they may not prove to be conducive to your survival.

A problem of shared values

IT IS TO be expected that most readers will look at the above sketch of the ways of our common, historical humanity and consider many of them backward, non-progressive and incompatible with modern ways of living. And they would be right, of course. Nevertheless, the discussion is worth considering, not because old traits must be emulated or readopted, but because they allow us to get at something else. You see, what if it turns out that your values, which you consider enlightened, progressive and above all yours, actually turn out to be the technosphere's values and are completely in line with the technosphere's own needs and motivations rather than with your own? And if you and the technosphere turn out to share the same set of values, then how on earth can you ever even hope to stand up in opposition to it? Stand up and do what—surrender?

Let's go down the list and give you ample opportunity to examine your own feelings.

Do you think that it's a good idea for people to generally live wherever their ancestors came from, to make good use of various physiological adaptations that they have developed over time, such as dark skin, or a stocky build and a generous layer of subcutaneous fat, or the ability to handle a certain endemic disease load? Perhaps you feel that this arrangement is too confining and limiting of individual freedom of movement and that people should instead be allowed to range over the entire planet as they do now. After all, if they spend most of their lives within a germ-free air-conditioned environment, what difference could their physiological adaptations possibly make? Well, they won't matter, until the technosphere goes away and takes the artificial environment with it. Then you would have stocky pale northerners stuck in the tropics, dying of heat exhaustion and sunburn, and lanky, dark-skinned people adapted to the southern deserts dying of frostbite and hypothermia in the snowy north.

On the other hand, cultural and ethnic diversity does seem like a winner—so liberal and progressive-sounding!—until you realize what it means to people who claim that it is their right to only deal with their own kind, except perhaps for a bit of trade, bride-snatching and the odd raiding party. Would you grant them that right, or would you rather perish in the futile attempt to force all of them to live harmoniously in a single "multicultural" society and send their children to school with the children of strangers whose cultures are incompatible with theirs and so on? If a certain tribe only wants to spend time with its own kind and is unwelcoming toward all outsiders, would you insist on starting a war with them, or would you consider just letting them be?

From the point of view of the technosphere, tribal behavior is decidedly suboptimal. The technosphere wants to deal with individuals because individual humans are weak and easy to manipulate and dominate. But once they combine in tight-knit, cohesive groups they become very strong and willful. A hundred or so people who hold down a patch of ground, have their own agenda and are ready to die for each other are certainly a force to be reckoned with and not at all copacetic with the technosphere's objective of complete domination and control over all living things.

But now consider what happens when the technosphere goes away, taking the police, the courts, the jails and all the rest with it. Would you prefer to be surrounded by strangers, any one of whom could at any time turn on you, a lone individual in a frightening and unfamiliar world, or would you rather be surrounded by people who are most like you, whose character is transparent to you, whom you know and trust and possibly even love, and who are willing to die for you, and you for them? The choice seems obvious.

Moving down the list… do you like traditional, strict gender roles and a clear separation of concerns between biological sexes, or do you believe in gender equality, equal rights, fluid gender roles, shared responsibilities for everything, complete acceptance of homosexuals and transgender individuals and so on? Again, the latter sound progressive, liberal, in some countries even patriotic, while the former sound decidedly old-fashioned and obsolete. You probably like the latter more than the former.

But what does the technosphere like better? Does it like it more when men behave like men and cultivate unquestioned, rock-solid male solidarity, while women are women and form a similar rock-solid tribal sisterhood, or would it prefer us to be maximally alienated from each other? Would it prefer the way men and women treat each other to be governed by a long-standing, inviolable tradition in which all are bound by the same unwritten, sometimes even unspoken code of conduct, or would it prefer that we weaken ourselves and our families with endless gender battles?

Perhaps the technosphere would prefer it if everyone were vaguely androgynous and sexually ambiguous, with meek, effete, ladylike men and women who are essentially emasculated men? After all, all today's men and women are ever required to do is push buttons and follow written instructions (until such time when they are duly replaced by algorithms and robots), and they can do these things well enough even if they are entirely unsexed. On the other hand why not indulge their sexual fantasies, no matter how perverse and bizarre? Why not make it socially acceptable to practice gay bestiality assisted by transsexual midgets? The more the merrier! The wider the spectrum of acceptable behavior, then the less people know what to expect of each other, the less likely it becomes for their interests and tastes to coincide, the more use-less they are to each other and, consequently, the easier they are to manipulate and control.

And would the technosphere prefer it if boys and girls had powerful role models in their fathers and mothers, respectively, and could learn all of the requisite survival skills simply by fol-lowing their parents around and assisting them in any way they could? Well, no, because this would make children strong-willed and independent-minded and that would get in the way of mak-ing them submit unquestioningly to being indoctrinated by licensed, credentialed educators and forced to memorize large amounts of useless trivia for the sake of passing standardized tests. (Such tests are poor educational tools, but they do estab-lish a performance standard for both students and teachers and so offer a wonderful way of controlling everyone.) No, it is much better from the technosphere's point of view for the parents to be confused or indifferent, generally passive, but eager to cooperate with educators for the sake of their children's educational success. After all, the technosphere wants your children to belong to it, not to you.

Next, let's consider the institution of marriage. Is the marriage ceremony a celebration of romantic love and a way of granting sexual relationships a bit of dignity? Is it acceptable for people to divorce and search for new, temporary love interests as soon as romantic love fades? Or should marriage be regarded as a lifelong contract that is based on feelings of duty to past and future gener-ations of your tribe, entered into with complete surrender of your individual interest for the sake of sustaining a greater whole?

Clearly, it is in the interest of the technosphere to make personal relationships as shallow, superficial and temporary as possible, so that the individual has no larger social entity to rely on. Strong extended families give individuals the ability to cultivate some amount of autonomy and freedom in group decision-making, and this is anathema to the technosphere, which wants to control everything through bureaucratic, technocratic management and supervision at the level of the individual. Weak families are also helpful in breaking the bond between generations, making the children more easily dominated by educators, more malleable and easier to control.

To this end, the extended family, with several generations living under one roof and with a single household budget—the bedrock of humanity since time immemorial—has been all but demolished, replaced by the nuclear family. And now it is the nuclear family that is being dismantled. According to the CDC, in the US in 2013 the percentage of out-of-wedlock births was 29.3 for whites, 54.2 for Hispanics and 71.4 for African Americans. Fathers have been made largely superfluous, and now it is the mothers' turn to be made redundant: because of the requirement to work, which makes no economic sense and is only made possible by subsidized daycare, children are brought up by low-paid strangers.

You may be justified in thinking that the modern social arrangement maximizes your personal freedom of choice and chances of finding sexual satisfaction. But what do you think will happen to nonexistent or weak nuclear families upon the disappearance of all the technosphere-provided services on which they depend? Chances are they will not last, because there is little substance to them beyond a living arrangement. When that living arrangement unravels, what is there to fall back on? At the other extreme, multigenerational extended families that see it as their sacred, inviolable duty to do everything possible to help their members even unto death should be able to do much better.

When it comes to freedom of movement, the modern arrangement attempts to break with the age-old human tendency to live out our days pretty much where we were born. The aspiration is to be mobile: to grow up in one place, be schooled in another and settle down in a third. Many people think nothing of switching houses, neighborhoods, towns, even countries as a side-effect of switching jobs. This arrangement is most useful from the point of view of the technosphere: labor flows to wherever it is needed. Nobody has any particular connection to a native piece of turf, and when it becomes trashed by economic development and turns into a barren, unsightly asphalt-and-concrete jungle they can simply move to someplace else that still needs more economic develop- ment. Since nobody has any particular connection to the people with whom they are temporarily thrown together, they have no opportunity to develop strong personal relationships that foster self-sufficiency and autonomy, making them easy to dominate and control.

But when the technosphere falls apart, this geographically mobile, rootless arrangement translates to being stranded among strangers. A population with a strong sense of rootedness—of being bound to a certain piece of terrain by ancestral lineages—will spontaneously form popular insurgencies to defend it against out- side threats. Members of a rootless population bereft of a profound sense of place will only stand up for themselves and perhaps for a few others, based on personal sympathies, and will be unable to spontaneously coalesce into a guerrilla fighting force.

Now we come to an even tougher subject: murder. The com- mandment "Thou shalt not kill" is a quirky one, given the quite tremendous amount of officially condoned murder that happens all the time. The all-time record in officially sanctioned murders was set in 2015, with China, Saudi Arabia and Pakistan at the top of the list and the US not far behind. Perhaps the commandment should be modified to "Thou shalt not kill unless so ordered by your superiors." That is, you aren't allowed to kill (unless it's in self-defense, or in some jurisdictions a crime of passion), but the technosphere is certainly allowed to kill, and you are allowed to kill on its behalf. Now, one odd thing about murder is that it tends to be extremely rare in places where there is little or no official law enforcement. This is because in such places a murder automatically leads to a blood feud, and relatives of the victim are more or less required to avenge it (unless the matter is resolved by paying blood money). But the technosphere certainly doesn't want you to take justice (or anything else, for that matter) into your own hands. What benefits the technosphere most is a high murder rate, to make people feel unsafe and clamor for more police protection because this makes them easier to control.

Lastly, we come to the matter of how we treat those with physical and mental abnormalities and those who, in politically correct language, are now to be referred to as "differently-abled." Of course, the enlightened, modern way is to deny that there is such a thing as "normal": all of us are on some sort of spectrum for all sorts of things—autism, obsessive-compulsive disorder, anxiety, depression, phobias, addiction, eating disorders, etc…. The goal is to allow all the various sufferers and the handicapped, regardless of the severity of handicap, to lead full, happy lives, with technology, both high and low, brought in to make this possible, from low-tech wheelchair ramps to high-tech suck-and-puff controlled wheelchairs and speech synthesizers.

A poster boy for this trend is Prof. Stephen Hawking, who held Sir Isaac Newton's chair at Oxford until his retirement in 2009. Hawking is almost completely paralyzed (and somewhat uncomfortable to look at) but is able to communicate profound cosmological thoughts through a speech synthesizer by moving his eyeballs about and twitching one of the few muscles, in his cheek, that is still wired up to his brain. Hawking recently said that we should take it easy on trashing the biosphere because we will still need it for a couple more centuries while we figure out how to get off this planet and colonize others. I suppose he hasn't heard that industrial civilization is almost over, but then he wouldn't be the only one.

It seems somewhat incongruous that Hawking long held Sir Isaac Newton's chair at Oxford—that same Newton one of whose

notebooks was written in Greek—classical Greek—that same classical Greece that idolized physical beauty and looked upon every
sort of deformity as an abomination. In classical Greece Hawking
would have been hidden away from the public, at the very least,
or carried to a forest and left to die, for fear of offending the gods
with his presence. But without classical Greek science both Newton and Hawking would have been professors of perfectly nothing,
because the entire modern scientific tradition got its start in that
one place and time.

It also seems somewhat incongruous that Hawking talks up
the idea of colonizing space, while Greek science would have had
nothing to do with anything so applied. It was a pure intellectual
pursuit in search of divine perfection. Sure, it was OK for Archime-
des to do a magic trick with mirrors to burn down the Roman fleet
in defense of his native Syracuse, but in a time of peace anything so
applied would have been considered undignified. Thus, to the classical Greeks, the technosphere would have been an impossibility,
while Hawking is its spokesman. From the classical Greek point of
view, Hawking is not just an abomination but also an embarrassment to science. But to us he is a hero because he has persevered in
spite of having a debilitating disease.

To many people today, our supportive treatment of those who
have problems—be it obesity or addiction or a little of each—is
a manifestation of our humanity. It is also legally required of us.
Fewer and fewer occupations, principally the police, firefighters,
paramedics and the military, require a fitness test; in all others the
handicapped have to be considered alongside able-bodied appli-
cants. It is a difficult subject because what is at issue is how much
of our compassion we are willing to sacrifice for the sake of our
safety and security.

But what does the technosphere want? It wants all of us to
be patients within the medical system. It has no standard of
health—just statistical measures of relative sickness. If all of us are
considered sick and in need of constant medical supervision, this
increases the amount of control it can exercise over us. If we are

weak, then this makes us more dependent on it and less able to get by without it. Just one family member who is in constant need of medical supervision is enough to make sure that the entire family can never risk losing access to medical treatments and will do whatever it takes to maintain that access.

But what will they do when the technosphere falls away, taking the medical system with it? Here, we have to put our compassion and part of our humanity aside. Anybody who isn't physically and mentally fit would automatically become a tremendous burden. Physically, anyone who can't walk and is too heavy to be carried by another person becomes a hindrance to movement. Mentally, anybody who suffers a nervous breakdown when the situation rapidly shifts for the worse can ruin the chances for everyone around them. As we work to shrink the technosphere, our ability to support those who cannot support themselves and are abjectly dependent on it will of necessity shrink too. It is an uncomfortable situation, but nobody has repealed natural selection and survival of the fittest. If we want to survive, we have to be fit and surround ourselves with those who are also fit.

If we value the exact same things that the technosphere finds useful for its own purposes, then our efforts to free ourselves from the technosphere's stranglehold are likely to fail. We should expect that many people will find themselves unwilling, unable or both, to change their values, even if they are capable of appreciating on an intellectual level that the values the technosphere has inculcated in them are maladaptive and will endanger their survival. On the other hand, there is no reason to think that all of the traditional, tribal human values are strictly necessary for survival; after all, human culture is extremely variable. Perhaps a community that embraces gender equality, is LGBT-friendly and is tolerant of handicaps and various human frailties and failings would be less efficient than a more traditional model, but this is just a guess, and what if it can find effective ways to compensate for this inefficiency? Ours is not to pre-judge; ours is to observe and to decide for ourselves. Perhaps modeling our society after Sparta would give us the best chances of surviving, but then how many of us would want to live like the Spartans?

How much do we need to compromise in deciding on how best to equip ourselves for the arduous tasks that lie ahead? Choosing tools and other bits of technology is relatively easy. There is a lot to learn, but all it takes is time and practice. It is much harder to choose the people with whom to surround ourselves, to support and to draw support from. There are plenty of things that can be compromised on relatively safely: physical beauty, youth, fashion sense, intellectual brilliance and an enlightened, progressive world-view and are, if you think about them hard enough, nonessential. But if you compromise on health, loyalty, common sense, adapt-ability, the ability to get along and that scarce but precious quality of rootedness—a strong sense of belonging to a place and a group of people—then the entire project is sure to be imperiled.

This choice is made even more difficult by the fact that virtually all of the people you can choose from are in one sense or another damaged or incomplete. Some have been too sheltered in their lives, while others not enough and have been traumatized. Some have been through various cycles of addiction, have hit bottom and recovered, while others have an unblemished history, but only because they were pampered by favorable circumstances and will instantly fall apart when stressed. How you choose may in the end say more about you than about those you choose. But choose you must, because the lone, supposedly rugged but in reality incredibly vulnerable individual does not stand a chance against the techno-sphere—for it could be said that humans are its second favorite food, after crude oil—and you will only be able to stand up to it as a group.

Why act now?

BUT WHAT, YOU may ask, is the urgency? After all, environmental degradation has been happening for a long time and will continue getting worse for centuries. Yes, the technosphere is becoming more and more invasive and oppressive all the time, but that's not

a new development either. Yes, nonrenewable resources are depleting, but they have been depleting ever since they were first tapped,
and pouring ever more money and energy into the extractive
industries seems to compensate for resource depletion for the
time being. Why is now—starting with when you finish reading
this book—the time for you to concentrate all of your efforts on
trying to shrink it? ✳ ✳

Well, here is the reason: the technosphere is terminally ill. As it
gets sicker and sicker, if we continue to depend on it, it will sicken
us as well. You see, the only reason it was able to continue to grow
is by chewing its way through larger and larger quantities of nonrenewable natural resources: oil, gas and coal, metal ores, fresh
water and arable land, phosphate for fertilizer and much else. And
now, it turns out, this trend is becoming impossible to sustain. A
thorough evaluation of the remaining supplies of nonrenewable
natural resources was conducted by Christopher Clugston and
detailed in his book *Scarcity: Humanity's Final Chapter.*[2] It is a convincing and compelling work, and it makes technophiles want to shoot
the messenger, because what it implies is that technological civilization is a suicide pact. And that is not a fact that technophiles are
able to take on board safely, because doing so would give them serious psychiatric problems.

Clugston examined statistics on nonrenewable natural
resources (NNRS) from USGS, EIA, BEA, BLS, the Federal Reserve,
CBO, FBI, IEA, the UN and the World Bank and concluded that
"absent some combination of immediate and drastic reductions
in our global NNR utilization levels... we will experience escalat-
ing international and intranational conflicts during the coming
decades over increasingly scarce NNRS, which will devolve into
global societal collapse, almost certainly by the year 2050." For
example, there remains only an eight-year supply of lithium—so
don't pin too many hopes on electric cars or portable computing

---

2   http://www.amazon.com/dp/1621412504

devices that use lithium-ion batteries. And there are only 15 years left of iron ore, the main ingredient in making steel. Most other resources are not far behind, with bauxite, used to make aluminum, one of the most plentiful, with a 40-year supply buffer.

Note that these supply problems are not something that is predicted to arrive in some remote, possibly fictional future; they are here today. Look at the roller-coaster ride of commodity prices so far this century, and a picture emerges of constant, permanent crisis. Commodity prices are, most of the time, either too high for the consumers to be able to afford the manufactured products or too low to allow the producers to extract the resources. As a result of this constant market whipsawing, strategic production planning becomes impaired, and losses mount for both producers and consumers and for the economy as a whole.

If this isn't the narrative you are used to hearing, then it's because there is a good reason for that. You see, whenever failing societies are forced to recognize that their problems are unsolvable, they tend to suffer something like a society-wide psychotic break and do all they can to persist in their conviction that everything is going to be all right. For example, when the train of the Soviet economy stopped moving and it became clear that only a rather different economy—one that did not have a role for the Soviet leadership—would make it possible to move forward, the leadership preferred to, figuratively speaking, draw the curtains, break out the vodka and the caviar, and pay flunkies to rock the train to pretend that it was still moving.

This is quite similar to what we are witnessing today: the inescapable reality of nonrenewable natural resource depletion has caused economic growth to slow and, in the more developed countries, to stall. In response, the monetary authorities have unleashed wave after wave of "stimulus": quantitative easing, zero interest rate policy (ZIRP) and now negative interest policy (NIRP). This has prevented the financial house of cards from collapsing outright, but monetary policy is unable to create super-giant oil fields full of light sweet crude or geologic strata of anthracite coal or high-grade hematite iron ore. Monetary policy isn't even able to prevent a market panic; all it can do is postpone it by some indeterminable but probably rather short period of time.

What this is, then, is the technosphere thrashing in its death agony. One moment the price of oil is too low, crushing the oil business; another moment it is too high, crushing the rest of the economy. When the supply and demand lines diverge and the price most oil consumers are able to pay without going bankrupt becomes lower than the price most oil producers have to charge to stay in business, it's effectively game over. In the meantime, we see fiscal austerity, increasing levels of financial instability, a continuously dropping labor participation rate, a shrinking middle class and more and more countries becoming failed states.

The urgency, then, comes from the need to avoid any of the following deadly eventualities:

The greatest danger, for most of us, is of **dying of withdrawal symptoms** as we lose access to technosphere's many products and services and can't come up with any way to compensate for their loss. For some people, such as those kept alive by industrial medicine, this is unavoidable. Others, especially if they are reasonably healthy and have access to some land, could shift to fishing, hunting, gathering and eventually growing their own food.

Next is the danger of **being stuck in the wrong place at the wrong time** and being crushed by the technosphere as it thrashes about or by being trapped underneath its lifeless, decaying hulk once it finally stops moving. The major population centers can be expected to be the most vulnerable. Deeply divided, internally conflicted societies held together by the overt threat of official violence euphemistically referred to as "law and order" are likely to suffer a great deal of looting, mayhem, rape and murder. And if the violence doesn't get you, high-density, built-up environments are not particularly survivable without functioning utilities and transportation networks.

At the other extreme, deeply rooted traditional societies that live largely off the land, do much of their own internal policing and are quick to take the law into their own hands in case of outside incursions should be able to do fine, but making peace with them and earning their respect and trust, which are essential if you wish to live alongside them, takes time, effort, some special talents and plenty of luck. It requires time—ten years or so on average—before the locals will accept you as one of their own. It also requires that you show a great deal of flexibility among people who are rather inflexible, and do so without losing face before them. In all, it is often easier to remain migratory or nomadic and be a welcome guest in several places than a permanent, unwelcome guest in just one.

The next-greatest risk is **getting poisoned or irradiated** by the various toxic and radioactive "gifts" of the technosphere—which will keep on giving long after it is gone. With regard to radiation, a good number to remember is the half-life of Plutonium-239, the isotope used to make nuclear weapons, which is over 24,000 years. Over 1,300 metric tons of it have been produced. In 24,000 years there will be only half as many tons of it left; in 48,000 years only a quarter as much. A few milligrams of it is a lethal dose. There are one billion milligrams in a metric ton, and the population of the Earth is just over seven billion. This means that there is much more than enough Plutonium-239 to kill all of us—perhaps as much as two grams per person—but only if each of us finds a way to receive a concentrated dose of it.

Plutonium-239 is but one example; there is also the problem of the much more plentiful spent nuclear fuel rods which are stored in pools of water at hundreds of nuclear power plants. The rods remain hot for a long time, and if cooling water isn't circulated and replenished using electric circulator pumps, it boils out, the rods catch on fire, cause hydrogen explosions, and plumes of radioactive dust enter the biosphere. (This exact scenario unfolded during the nuclear disaster at Fukushima Daiichi nuclear power plant in Japan in March of 2011.)

If some substantial part of the nuclear stockpile gradually becomes evenly dispersed throughout the biosphere—the oceans, the Earth's crust—as it most certainly will after a few tens of thousands of years of neglect, then, should any humans still be around, few of them will live long enough to reproduce because of high rates of cancer, and those that do will give birth to many children who will be nonviable because of birth defects. We have already started seeing signs of this in places that were bombed by the US or by NATO using depleted uranium ordnance—Serbia/Kosovo, Basra and Fallujah in Iraq, and elsewhere. High rates of cancer and birth defects do not necessarily spell extinction, at least not immediately, provided enough of the women have lots of children starting at a young age. (which, of course, is not happening)

But there is no reason to think that toxic and radioactive materials will be evenly dispersed, at least in the near future, and so the task of survival requires identifying those locations which are particularly unsafe, the better to avoid an encounter with a lethal dose. Since radiation cannot be perceived by our senses, without a Geiger counter you would be driving blindfolded, and it takes a good deal of scientific knowledge and engineering know-how to make one from scratch using artisanal methods. The situation is only slightly better with regard to toxic chemicals, because here our senses, if we are well-attuned to our environment, do serve as a rough guide: water that has healthy-looking plants and animals in it obviously isn't killing them while a body of water that is perfectly clear and transparent may be poisoned; an oily film on the water may indicate proximity to a hydrofractured oil and gas well that was hastily capped but has since started leaking toxic, possibly radioactive substances and is likely to continue doing so for decades. Fruits, berries and mushrooms that have an unusual, metallic taste should be avoided. There will be much to learn—not necessarily through science, but by collecting anecdotal evidence and by evolving a system of taboos—the way we hominids have managed to survive for millions of years.

Admittedly, none of these survival scenarios sounds all that happy. But we have foolishly allowed the technosphere to make our bed, and now we will have to sleep in it. Yes, the changes we need to make are, at the very least, uncomfortable: we have to break habits, we have to learn to do without luxuries and deprive ourselves of comforts; we have to change our location, acquire new skills, make new friends and adopt a different culture and a different outlook—ones predicated not on achieving success within a successful society but on survival on the edges of a failed one.

3

APPROACHES AND DEPARTURES

THIS BOOK IS by no means the first critique of technology to
have ever been written. Technology has been roundly criticized
from numerous angles and mindsets: from the romantic and the
aesthetic (which often find technology to be less than entirely
pleasing) to the political (from which it is often seen as a tool of
capitalist and/or communist exploitation) to the sociological
(which sees in it a source of social inequality). There is also no
shortage of critics of separate industries, with the nuclear power
industry, the fossil fuel industry and the medical industry number-
ing among the favorite targets.

But this is the first book to explicitly define the technosphere
as an organism and an emergent intelligence that has enslaved and
is destroying the biosphere—and us with it. Nor has there been
an effort to offer a comprehensive, constructive plan for how to
get a grip on it and make it serve our purposes rather than allow
ourselves to blindly follow in the wake of its own pathological, ultimately suicidal course. Nor was this book in any way inspired by
any of the existing critiques of technology, most of which, upon a
cursory examination, have turned out to be either toothless, short
on solutions or both. Lastly, it is not even exactly a "critique of
technology"—any more than a book on gardening is a "critique of
plants." Nevertheless, it seems proper to mention two of the most

far-reaching critiques of technology to date and to give them credit
where credit is due.

Jacques Ellul

THE ONE THINKER who came closest to defining the technosphere was Jacques Ellul, author of *The Technological Society*,[1] in
which he offered a critique of industrial technology as an all-encompassing system rather than a mere set of instrumentalities
mediating human interactions with nature. His prose is dense and
to some, especially in English translation, almost impenetrable, but
for those who persevere his analysis is quite good.

However, he fails to provide a way out. This may have had to
do with his Christian outlook. To him, the existence of the eternal
human soul and an anthropocentric spiritual realm could not possibly be regarded as just another bit of technology. Thus, he chose
to examine the technosphere as a social phenomenon, ignoring the
individual whose soul was held harmless within the Church and
her sacraments awaiting its salvation.

To my mind, organized religions are social machines par excellence, and thus they are a form of technology (social machines are
covered in detail later in this book). Religions are in many ways
similar to other social machines, such as corporations and government agencies. They have their own codices of laws, their own
bureaucracies and their own mechanisms for exacting compliance. Like other social machines, they strive to limit the range of
individual action. But unlike other social machines, they declare
themselves to be above and beyond the reach of physical law. They
demand of their followers a theatrical suspension of disbelief with
regard to impossibilities and things invisible, but then they extend

---

1   Jacques Ellul, *The Technological Society*, trans. John Wilkinson (New York: Vintage/
Alfred A. Knopf, 1964), originally published as *La Technique ou l'enjeu du siècle* (Paris:
Armand Colin, 1954).

such theatricalities to matters of life and death. I see Ellul's inability to see his Christianity as a form of technology as a huge blind spot. I suspect that it is this that explains his inability to move from analysis to application. "It is not possible for me to treat the individual sphere," he conceded bleakly. Apparently, we are expected to meekly submit to the technosphere in this world, awaiting our salvation in the notional next.

But Ellul did come very close to defining the technosphere. This was a major achievement, and it is therefore surprising how little of his thinking echoes in the works of later critics of technology. Many other writers do mention him, and perhaps they have even thumbed through his treatise, but apparently they were somehow unable to absorb the full import of his discovery, instead concentrating on this or that element of technology that to them didn't seem quite right and called for some critiquing. Perhaps this has something to do with Ellul's choice of terminology, or perhaps something got lost in translation. To him, the technosphere was simply la technique—or, sometimes, les techniques—and he overloaded these terms mercilessly. The technosphere is not a technique or a technology. Unlike la technique, it is not an intellectual or a metaphysical construct; it is made of concrete, steel, plate glass, fluorescent ceiling lights and synthetic wall-to-wall carpeting, just as the biosphere is made of plants and animals, soil, air and water.

Fancy intellectualizing aside, the technosphere is not a metaphysical construct but a physical phenomenon. I remember one of the moments when it started to come together for me. I was sailing down the coast toward Boston Harbor. After a few days of nothing but an empty horizon and a few sea birds for company, my senses were finely tuned to the natural environment and easily insulted by first contact with the unnatural. First I floated over the 24-foot diameter outflow tunnel which discharges treated sewage from Deer Island Sewage Treatment Plant into Massachusetts Bay at a 100-foot depth. Then I meandered into the transatlantic approach path of Logan International Airport, with each jet sending a sooty heat wave blasting down on me, reeking of unburned kerosene as it shrieked past. Finally, the broken-toothed jaw of the Boston skyline hove into view, complete with the ever-present pillow of purple haze sitting on top of it. No, you don't need keen intellectual abilities to discover the technosphere; your senses of sight, hearing and especially smell are quite enough.

To me, intellectualizing is not quite as good as firsthand experience, but it can be quite impressive. Here are some passages from Ellul's *The Technological Society* which I have found to be important, with his choice of words sometimes replaced by mine. I don't believe that my replacements alter his meaning in any significant way. Please judge for yourself, but it seems to me that he got quite close to what I have been trying to get at here.

> [The technosphere] cannot be otherwise than totalitarian. It can be truly efficient and scientific only if it absorbs an enormous number of phenomena and brings into play the maximum of data. In order to coordinate and exploit synthetically, [the technosphere] must be brought to bear on the great masses in every area. But the existence of [the technosphere] in every area leads to monopoly...
>
> The individual in contact with [the technosphere] loses his social and community sense as the frameworks in which he has been operating disintegrate under the influence of [the technosphere]. This fact is established beyond question by the disappearance of responsibilities, functional autonomies, and social spontaneities, the absence of contact between the technical and the human environment, and so forth...
>
> [The technosphere] makes its sociological compost pile where it does not find one already made. And it possesses sufficient power and efficiency today to succeed. Before long, it will produce everywhere that clear technical consciousness which is the easiest of its creations to bring about, and which man falls in with so willingly. The world that [the technosphere] creates cannot be

any other than that which was favorable to it from the very beginning. In spite of all the men of good will, all the optimists, all the doers of history, the civilizations of the world are being ringed about with a band of steel. We in the West became familiar with this iron constraint in the nineteenth century. Now [the technosphere] is mechanically reproducing it everywhere as necessary to its existence. What force could prevent [the technosphere] from so acting, or make it be otherwise than it is? (125-127)

Ellul was also quite on target with regard to what I call the *technologizing effect* of the technosphere on all human endeavors. Here, his relentless use of the term *la technique* requires less semantic stretching (it's hard to think of a technique as "totalitarian"—that's more of a property of a system that *uses* that technique), but the word he really seems to be reaching for is *technology*.

[Technology] is the main preoccupation of our time; in every field men seek to find the most efficient method.

1. Economic [technology] is almost entirely subordinated to production, and ranges from the organization of labor to economic planning. This [technology] differs from the others in its object and goal. But its problems are the same as those of all other [technologies].
2. The [technology] of organization concerns the great masses and applies not only to commercial or industrial affairs of magnitude (coming, consequently, under the jurisdiction of the economic) but also to states and to administration and police power. This organizational [technology] is also applied to warfare and insures the power of an army at least as much as its weapons. Everything in the legal field also depends on organizational [technology].
3. Human [technology] takes various forms, ranging all the way from medicine and genetics to propaganda (pedagogical [technologies], vocational guidance, publicity, etc.). Here man himself becomes the object of [technology]. (20-22)

Finally, Ellul was quite prescient (he wrote his treatise over six decades ago) of the technosphere's runaway mode and destructive nature. He realized that it is not under anyone's control, that it claims its own prerogatives and pursues its own aims by perverting human values. It is beyond morality, beyond conscience. Human interests and concerns are entirely incidental to its pursuit of dominance and are used as mere instruments to motivate and incentivize humans to serve it.

Modern society is, in fact, conducted on the basis of purely technical considerations. But when men found themselves going counter to the human factor, they reintroduced—and in an absurd way—all manner of moral theories related to the rights of man, the League of Nations, liberty, justice... When these moral flourishes overly encumber technical progress, they are discarded—more or less speedily, with more or less ceremony, but with determination nonetheless. This is the state we are in today.

Technical progress today is no longer conditioned by anything other than its own calculus of efficiency. The search is no longer personal, experimental, workmanlike; it is abstract, mathematical, and industrial. This does not mean that the individual no longer participates. On the contrary, progress is made only after innumerable individual experiments. But the individual participates only to the degree that he is subordinate to the search for efficiency, to the degree that he resists all the currents today considered secondary, such as aesthetics, ethics, imagination. Insofar as the individual represents this abstract tendency, he is permitted to participate in technical creation, which is increasingly independent of him and increasingly linked to its own mathematical law. (74)

...In our civilization [technology] is in no way limited. It has been extended to all spheres and encompasses every activity, including human activities. It has led to a multiplication of means without limit. It has perfected indefinitely the instruments available to man, and put at his disposal an almost limitless variety of intermediaries and auxiliaries. [The technosphere] has been extended geographically so that it covers the whole earth. It is evolving with a rapidity disconcerting not only to the man in the street but to the technician himself. It poses problems which recur endlessly and ever more acutely in human social groups. Moreover, [technology] has become objective and is transmitted like a physical thing; it leads thereby to a certain unity of Civilization, regardless of the environment or the country in which it operates. (78)

Modern men are so enthusiastic about [technology], so assured of its superiority, so immersed in the technical milieu, that without exception they are oriented toward technical progress. They all work at it, and in every profession or trade everyone seeks to introduce technical improvement. Essentially, [technology] progresses as a result of this common effort. Technical progress and common human effort come to the same thing. (85)

The human being is delivered helpless, in respect to life's most important and most trivial affairs, to a power which is in no sense under his control. For there can be no question today of man's controlling the milk he drinks or the bread he eats, any more than of his controlling his government. (107)

I discovered Ellul only after I had started writing this book, when I undertook a search for any precursors for my way of thinking among those who have been classified as critics of technology. I was quite amazed to find such a kindred spirit across so many decades. I was also surprised that those that followed him not only failed to build on his ideas, but failed to make much progress at all. Lots of smart people made an effort, but they all seemed to lack the unifying idea and were like blind people groping various parts of an elephant, not realizing that it is an elephant—the elephant I chose to call "the technosphere."

Ted Kaczynski

I LEARNED OF another kindred spirit thanks to the presentation Albert Bates gave at the 3rd annual Age of Limits Conference, which was held at the Four Quarters Interfaith Sanctuary in Artemas, Pennsylvania, in 2013. His talk was about Ted Kaczynski, whom the FBI called the Unabomber (for "University and Airline Bomber"). Most people who have heard of Kaczynski think of him the way he was portrayed in the media—as a paranoid schizophrenic terrorist. The story Bates told us at the conference was in stark contrast to this slanted, superficial account. (I excerpted the following extended quote from Albert's blog.)[2]

Kaczynski was born and raised in Chicago. He was accepted to Harvard at 15. He got his PhD in mathematics at University of Michigan and became an assistant professor at [the University of California at] Berkeley in 1967, at age 25. [He] was an extraordinary genius whose sense of human dignity was profoundly altered, at age 16, by being secretly made the subject of an MKULTRA mind control experiment while a child prodigy undergrad at Harvard. He was a casualty of the Cold War. During the test, gifted students who volunteered for the program were taken into a room and connected to electrodes that monitored their physiological reactions while facing bright lights and a one-way mirror. Then they were brutally confronted with their inner demons, that they had provided the interrogators during months of screening tests. LSD or other drugs may have played a role. This horrific experience fostered an abiding animus, not just in Ted Kaczynski, but in all the subjects, towards the secretive security state....

Kaczynski's students at Michigan all said he was an excellent instructor, but the opposite was reported at Berkeley. His goal by then was not to teach, but to save up to get a cabin in Montana.... Kaczynski built the cabin himself, lived with very little money, and without electricity, telephone or running water. He studied tracking and edible plant identification and gained primitive skills. The ultimate catalyst that drove him to begin his bombings was when he went out for a walk to one of his favorite wild spots,

2   http://PeakSurfer.blogspot.com

only to find that it had been destroyed and replaced with a Forest Service road. He stopped studying nature and began studying bomb-making.

Kaczynski: "As I see it, I don't think there is any controlled or planned way in which we can dismantle the industrial system. I think that the only way we will get rid of it is if it breaks down and collapses."

From 1978 to 1995, Kaczynski sent 16 bombs to targets including universities and airlines, killing three people and injuring 23. Kaczynski is serving life without possibility of parole in the Florence, Colorado Supermax. He is an active writer, and his current writings are stored at the University of Michigan Special Collections Library. They are embargoed until 2049. His Montana cabin, transplanted board for board, is on display at the Newseum in Washington, D.C. On May 24, 2012, Kaczynski submitted his current information to the Harvard University alumni association. He listed his eight life sentences as achievements, his current occupation as prisoner, and his current address as No. 04475-046, US Penitentiary—Max, P.O. Box 8500, Florence, CO 81226-8500.

It stands to reason that Kaczynski's brilliant intelligence, conditioned by the personal experience of, essentially, repeatedly undergoing mental torture while under the influence of hallucinogenic drugs, made obvious to him a reality that others could not see. He was being victimized by the bit of the technosphere that is by far the nastiest: the part that seeks to make humans 100 percent controllable, as if they were robots, by breaking down everything about them that makes them human. I am certain that he caught a glimpse of what the technosphere really is: a single, unified, global, controlling, growing, destructive entity, existing beyond human reason or morality, which must be stopped no matter the cost.

Although society convicted him as a violent criminal, his final score of three dead and 23 injured is paltry compared to the scores of dead and injured racked up by, for instance, any of the recent

US presidents during their various wars of choice, which number in the hundreds of thousands dead and millions injured. Yet none of them has been convicted of murder because they were arguably simply doing their job, and killing and maiming lots of people (who are euphemistically referred to as "collateral damage") is presumably part of that job. Nor have any of those who carried out their murderous orders been brought to any form of justice. Kaczynski violated the real First Commandment: "Thou shalt not kill *unless so ordered.*" But what if he saw it fit to assume command and gave *himself* the order to kill as part of simply doing *his* job—the very important job of drawing attention to the technosphere with the eventual goal of destroying it? What if following the morality of the powerful is nothing more than a way to secure for yourself a place among the losers? What if, as Nietzsche said, "Morality is the herd instinct of the individual"?[3]

In his *Industrial Society and Its Future* (1995),[4] which is often referred to as the *Unabomber Manifesto*, Kaczynski wrote:

We therefore advocate a revolution against the industrial system. This revolution may or may not make use of violence; it may be sudden or it may be a relatively gradual process spanning a few decades. We can't predict any of that.

In order to get our message before the public with some chance of making a lasting impression, we've had to kill people.

But first Kaczynski did everything he could, peacefully. He dropped out of the mainstream, he built a cabin in the wilderness, and he lived there in accordance with his principles. *If everyone lived like Ted, there would be no technosphere for me to write about.* But then the Forest Service encroached on his habitat, destroying a part of it, forcing him to act. The actions he took were symbolic and his

---

3   The *Joyful Wisdom* (1882)

4   Numerous online sources, including http://editions-hache.com/essais/pdf/ kaczynski2.pdf, accessed June 2016.

bomb-making operation was artisanal rather than industrial. His bombs were a kinetic way to send an intellectual message. In this he succeeded, and it is a testament to his effectiveness as a terrorist that his manifesto is being assigned as reading in university class-rooms and that his story is being retold on the pages of this book.

Kaczinsky echoed Ellul's statement that "[The technosphere] cannot be otherwise than totalitarian." But he went further, explaining how the technosphere's prerogatives have nothing to do with helping humanity but everything to do with exploiting humanity to perpetuate the machine itself, to humanity's own detriment.

> ... A "free" man is essentially an element of a social machine and has only a certain set of prescribed and delimited freedoms; free-doms that are designed to serve the needs of the social machine more than those of the individual.
>
> The system does not and cannot exist to satisfy human needs. Instead, it is human behavior that has to be modified to fit the needs of the system. This has nothing to do with the political or social ideology that may pretend to guide the technological system. It is not the fault of capitalism and it is not the fault of socialism. It is the fault of technology, because the system is guided not by ideology but by technical necessity.
>
> Need more technical personnel? A chorus of voices exhorts kids to study science. No one stops to ask whether it is inhumane to force adolescents to spend the bulk of their time studying sub-jects most of them hate. When skilled workers are put out of a job by technical advances and have to undergo "retraining," no one asks whether it is humiliating for them to be pushed around in this way. It is simply taken for granted that everyone must bow to technical necessity.
>
> ... All these technical advances taken together have created a world in which the average man's fate is no longer in his own hands or in the hands of his neighbors and friends, but in those of politicians, corporation executives and remote, anonymous

technicians and bureaucrats whom he as an individual has no power to influence.

Once a technical innovation has been introduced, people usually become dependent on it, so that they can never again do without it, unless it is replaced by some still more advanced innovation. Not only do people become dependent as individuals on a new item of technology, but, even more, the system as a whole becomes dependent on it.

Those who want to protect freedom are overwhelmed by the sheer number of new attacks and the rapidity with which they develop, hence they become apathetic and no longer resist. To fight each of the threats separately would be futile. Success can be hoped for only by fighting the technological system as a whole; but that is revolution, not reform.

And there we have it: "revolution, not reform." But who would be its revolutionaries? It would appear that at the moment there is just one revolutionary: Kaczynski himself. But since he happens to be a permanent resident of a maximum-security prison and even his newer writings are embargoed until a date far in the future, he won't be available to move the process along. One manifesto, no matter how well-written or well-read, does not a revolution make: it takes some revolutionaries to make a revolution.

Can the revolutionary task be delegated to some part of the existing political spectrum? Kaczynski is quite scathing in his criticism of the liberal left:

When we speak of leftists in this article we have in mind mainly socialists, collectivists, "politically correct" types, feminists, gay and disability activists, animal rights activists and the like.

The two psychological tendencies that underlie modern leftism we call "feelings of inferiority" and "oversocialization."

The moral code of our society is so demanding that no one can think, feel and act in a completely moral way. For example, we are not supposed to hate anyone, yet almost everyone hates somebody at some time or other, whether he admits it to himself or not. Some people are so highly socialized that the attempt to think, feel and act morally imposes a severe burden on them. In order to avoid feelings of guilt, they continually have to deceive themselves about their own motives and find moral explanations for feelings and actions that in reality have a non-moral origin. We use the term "oversocialized" to describe such people.

Oversocialization can lead to low self-esteem, a sense of powerlessness, defeatism, guilt, etc. One of the most important means by which our society socializes children is by making them feel ashamed of behavior or speech that is contrary to society's expectations.

… Oversocialization is among the more serious cruelties that human beings inflict on one another."

The guardians of political correctness (mostly white, upper-middle-class heterosexuals) identify with groups of people they see as weak or inferior while denying (even to themselves) that they consider them to be such.

Among "leftist issues," Kaczynski counts "racial equality, equality of the sexes, helping poor people, peace as opposed to war, nonviolence generally, freedom of expression, kindness to animals." Please note that "destroying the technosphere before it destroys us" is nowhere on this list.

So much for the liberals. As far as the conservatives, he dispatches them with a single mordant remark:

The conservatives are fools: they whine about the decay of traditional values, yet they enthusiastically support technological progress and economic growth. Apparently it never occurs to them that you can't make rapid, drastic changes in the technology and the economy of a society without causing rapid changes in all other aspects of the society as well, and that such rapid changes inevitably break down traditional values.

Although Kaczynski does not state this explicitly, the obvious implication is that there is nobody within the political spectrum of bourgeois democracy who could lead the charge to dethrone the technosphere from its position of global dominance. With him in jail, prospects for revolutionary change seem bleak.

But he does provide some hints of what such a movement might center on, and his suggestion is as brilliant as it is simple. We must focus on human needs—our needs, the ones the technosphere is failing to meet. In catering to these needs, we can outcompete the technosphere hands down because we can provide something that the technosphere cannot provide without undermining itself: namely, **autonomy, self-sufficiency and freedom**.

> ... Most people need a greater or lesser degree of autonomy in working toward their goals. Their efforts must be undertaken on their own initiative and must be under their own direction and control. Yet most people do not have to exert this initiative, direction and control as single individuals. It is usually enough to act as a member of a *small* group.

Being deprived of a sense of autonomy leads to disempowerment, whose symptoms are "boredom, demoralization, low self-esteem, inferiority feelings, defeatism, depression, anxiety, guilt, frustration, hostility, spouse or child abuse, insatiable hedonism, abnormal sexual behavior, sleep disorders, eating disorders, etc." But such disempowerment is essential because

> ... A technological society *has to* weaken family ties and local communities if it is to function efficiently. In modern society an individual's loyalty must be first to the system and only secondarily to a small-scale community, because if the internal loyalties of small-scale communities were stronger than loyalty to the system, such communities would pursue their own advantage at the expense of the system.

Freedom means being in control (either as an individual or as a member of a small group) of the life-and-death issues of one's existence: food, clothing, shelter and defense against whatever threats there may be in one's environment. Freedom means having power; not the power to control other people but the power to control the circumstances of one's own life.

The positive ideal that we propose is Nature. That is, wild nature: those aspects of the functioning of the Earth and its living things that are independent of human management and free of human interference and control. And with wild nature we include human nature, by which we mean those aspects of the functioning of the human individual that are not subject to regulation by organized society but are products of chance, or free will, or God (depending on your religious or philosophical opinions).

Small-scale technology is technology that can be used by small-scale communities without outside assistance. Organization-dependent technology is technology that depends on large-scale social organization. We are aware of no significant cases of regression in small-scale technology. But organization-dependent technology *does* regress when the social organization on which it depends breaks down.

Here then is the seed of a solution: pursue a strategy of forming small, autonomous communities making use of small-scale technologies that are not organization-dependent and do not regress. By doing so, we can starve the technosphere of labor and resources, *forcing* its organization-dependent technologies to regress. This will force it to relinquish control, allowing both cultural and economic life to become intensely local, deeply rooted in the surrounding biosphere and fully invested in its health. This strategy needs an ideology to frame it, and this book is part of the effort to develop just such an ideology—which is precisely what Kaczynski called for:

It is necessary to develop and propagate an ideology that opposes technology and the industrial system. Such an ideology can become the basis for a revolution against industrial society if and when the system becomes sufficiently weakened. And such an ideology will help to assure that, if and when industrial society breaks down, its remnants will be smashed beyond repair, so that the system cannot be reconstituted.

Kaczynski makes it clear that even if all goes as well as can be expected with this program, there will unavoidably be much death and suffering:

> If the breakdown is sudden, many people will die, since the world's population has become so overblown that it cannot even feed itself any longer without advanced technology. Even if the breakdown is gradual enough so that reduction of the population can occur more through lowering of the birth rate than through elevation of the death rate, the process of deindustrialization probably will be very chaotic and involve much suffering.

But then, what other choices are there? Moreover, what is the point of dwelling on this? If discussing it leads us to develop an exaggerated sense of responsibility for what happens, then we would have to accept that responsibility without gaining any ability to control what happens, because it is the technosphere that is in control, not us. Responsibility without control causes stress. But who needs extra stress? In a harm/benefit analysis (which is the topic of the next chapter) this way of thinking is harmful (because it causes stress) and has no benefits (because there is nothing we can reasonably hope to achieve by thinking this way). Therefore it belongs at the very bottom of the list, among various other things that we shouldn't bother doing.

It is somewhat ironic that Ted should exhibit such scruples, given that his chosen technique of bringing the problem to the world's attention was by killing and maiming people. Perhaps he is not such a revolutionary after all; were he a true revolutionary, Field Commander Kaczynski would simply have exclaimed, "¡Que será, será!" and let the human chips fall wherever they may. But if we too have such scruples, then, following Ted's logic, we have to consider which category of fools we are part of: liberal hypocrites, whose feigned guilt leads them to pretend to care about those they secretly consider inferior, or conservative fools, who continue believing in progress even as it destroys everything they wish to conserve. Perhaps we should set aside misplaced feelings of guilt for things we can never hope to control and instead cultivate a *healthy* sense of guilt—over our own inaction. Although his methods were unsound, Ted *did* something. I wrote this book. What have *you* done?

4

HARM/BENEFIT ANALYSIS

WE ARE NOW READY TO define the main strategy for shrinking the technosphere down to size, to deprive it of the ability to exercise its own prerogatives to grow, control, dominate and ultimately destroy the biosphere. But living without any technology whatsoever is out of the question: as I have explained, humans, and our hominid ancestors before us, have been tool-makers and tool-users for close to three million years. On the other hand, living with the technosphere, while it depletes and destroys the biosphere taking us with it, is also out of the question. And so, a third way must be found—a way that allows us to select specific technologies that we wish to preserve while rejecting the rest. A way must be found to consciously reduce technological complexity—the number of technical elements we allow within our environment—while increasing biological complexity—the number of other species we cultivate and support within our local patch of the biosphere.

Following this approach should bring good (or at least, better) results for you and yours. How it impacts the rest of humanity is out of scope; nobody has ever nominated either you or me for world leader. As the Chinese saying goes, "Let every man sweep the snow from before his own doors and not trouble himself about the frost on his neighbor's tiles."[1] But provided enough people see their own personal advantage in following this strategy and do follow it, the technosphere's grip on us will weaken, our grip on it will strengthen, and over time we will shrink the technosphere down to a size at which it will no longer pose much of a threat. It will no longer control us; instead, we will control it, and use it for one thing only: to provide us with tools that will help us lead autonomous, self-sufficient, free lives in the midst of nature, unencumbered by excessive organizational or technological complexity.

As explained in the previous chapter, according to Ted Kaczynski, we need to reject organization-dependent technologies that tie us into the technosphere and cultivate organization-independent ones. Easier said than done! It implies eliminating just about everything that makes it possible for people to survive. It implies living without electricity—not even off-grid systems that use batteries, photovoltaic cells and small-scale wind generators, because the supply chain for the necessary components spans the entire planet. It means living without pumped water, because pumps, pipes and valves are all manufactured products. It means living without electronics of any kind, since the electronics industry is globally integrated. No internet; no vaccinations; no cosmetic dentistry; no eyeglasses; no antibiotics or painkillers... Nothing that's mass-produced... It means living off the land using crude tools you can fashion yourself in a primitive smithy using salvaged metal. Yes, it is possible, and even today there are a few people left here and there who can live this way. Very few other people would ever settle for that!

Sorry, Ted, but we need a better metric on which to base our decisions than simply sorting technologies into organization-dependent and organization-independent and depriving ourselves of all the organization-dependent ones. So how about we do this instead: define a reasonably complete list of positive and negative aspects of technology, and then select which technologies we will

---

I   Brown, Brian: *The Wisdom of the Chinese: Their Philosophy in Sayings and Proverbs* (Charleston: Nabu, 2009), 185.

use in order to maximize the benefit while minimizing the harm. At the very least, it will give all of us a place to start; at the very best, it will shrink the technosphere to a point where it no longer poses much of a danger.

Calculating the harm/benefit ratio

UNLIKE THE EXTREMIST approach outlined above, this is a perfectly copacetic, constructive initiative, but I believe that it can achieve the same end result, albeit more gradually. You see, the harm/benefit analysis is designed in a way that maximizes technology's benefit to *us* while minimizing technology's harm to *us—not to the technosphere*. And I would conjecture that, based on which aspects of technology we regard as positive or negative, we can structure the process so that whatever helps us more or less automatically hurts the technosphere.

Shown in the table that follows are 32 aspects of technology, in no particular order, which, for each technology, take a value somewhere between harmful and beneficial.

To analyze a particular technology, decide for each of these aspects whether the technology is harmful or beneficial and assign a score of 1 for either harm h or benefit b.[2] To determine its Harm/ Benefit Ratio (HBR) tally each column and divide the total harm ($H=\sum h$) by the total benefit ($B=\sum b$):

$$HBR = \frac{H}{B} = \frac{\sum h}{\sum b}$$

Note that there is nothing magical about the 32 aspects of technology listed below, and you can modify this list or come up with one entirely your own (although sticking to this common list will make it easier for you to compare notes with other people and to reach joint decisions). It is just a way of evaluating the pros and

---

2   Where needed, a more granular analysis is also possible, say, by ranking every h and every b on a scale from 1 to 10.

	Harmful	Beneficial
1	toxic/radioactive	inert/biodegradable/edible
2	disposable	maintainable
3	mandatory	optional
4	limited useful life	unlimited useful life
5	fosters dependence	fosters autonomy
6	standardized	custom
7	expensive	free
8	obsolescent	perpetual
9	single-purpose	multiple-purpose
10	depletes resources	conserves resources
11	artificial	natural
12	synthetic	organic
13	industrial	artisanal
14	limits options	opens up possibilities
15	transnational	local
16	requires specialists	requires generalists
17	classifiable	unclassifiable
18	transparent	ambiguous/opaque[3]
19	individual use	community use
20	new	(re)used
21	consumer grade	commercial/military grade
22	retail	wholesale
23	packaged goods	bulk goods
24	rarely used	frequently used
25	networked	standalone/peer-to-peer
26	externally powered	unpowered/self-powered
27	automatic	manual
28	branded	generic
29	proprietary	open-source
30	licensed/registered	anonymous
31	requires energy	requires skill
32	individual effort	team effort

3   This may seem backwards, but it isn't. In many situations transparency invites
    exploitation by outsiders, while ambiguity and opaqueness require local
    knowledge and subjective judgment.

cons of technology but with a particular view in mind: what is considered beneficial is that which is beneficial to *you*, within *your* local environment, human or natural, which makes you autonomous, self-sufficient and free. And what is considered harmful is what disrupts the natural environment while depriving you of autonomy, self-sufficiency and freedom, forcing you to relinquish control to impersonal, remote, nonhuman entities.

There are some limitations to an approach which treats the set of harmful and beneficial features of each piece of technology as a flat list, because a more complex internal structure may be present: many technologies are interdependent. There are some technologies that we could ignore if only we *had* certain other ones, while there are other technologies that we could ignore if only we could *do away with* certain other ones. There are also technologies which we cannot ignore, and we should therefore pay most careful attention to them.

Anti-technology technologies

SOME TECHNOLOGIES ACT in opposition to each other, and one of them can be used to offset the other:

- Offensive vs. defensive weapons
- Law enforcement technologies vs. technologies for defying and defeating them
- Intellectual property rights enforcement technologies vs. technologies that allow you to ignore intellectual property rights
- Technologies for debt collection vs. technologies that allow you to repudiate your debts
- Technologies for prosecuting trespassers and squatters vs. technologies trespassers and squatters can use to avoid detection and prosecution, or to take control of property by adverse possession
- Technologies for objectively identifying people, animals and objects in order to control them vs. technologies for rendering them unclassifiable, unrecognizable, undetectable and anonymous

- Technologies for directing public discourse along rationalist, reductionist, materialist lines vs. technologies for negating these efforts through understandings privately shared within a group that are subjective, intuitive and mystical

When it comes to such negating technologies, it is not sufficient to examine them from the point of view of their harm/benefit analysis taken separately. Instead, we should count the harm they cause to the technologies they negate as a benefit. Given a technology T and an anti-technology t, the harm t causes to T is added to t's benefit:

$$HBR_t = \frac{H_t}{B_t + H_T}$$

The best anti-technologies provide a tremendous cost advantage relative to the technologies they negate. The original *Saboteurs* jammed expensive machinery using cheap wooden clogs (*sabots* in French). For a modern example, a simple infrared LED worn on your hat, invisible to the naked eye and put together using a few dollars' worth of widely available components, can temporarily blind video surveillance systems costing thousands of dollars. A spark gap generator, similarly cheap, simple and crude, can disable advanced radio communications systems.

Less crude but still remarkably cost effective are anti-technologies such as electronic warfare equipment. A box of Russian electronics can remotely shut down a US destroyer or aircraft carrier (as was demonstrated in the Black Sea, so it has been reported, when the Russians consolidated their right to maintain a port in Crimea following the destabilization of the Ukraine). A thumb drive, in the right hands, can be used to breach the defenses of a multibillion-dollar network security system. This is how the Stuxnet computer worm was able to disable scores of Iranian centrifuges used for uranium enrichment.

And the worst anti-technologies, barely deserving of the name, are replacement technologies that are better than the technologies they replace but still not good enough. A good example is replacing nuclear power plants with wind generators and photovoltaic panels. While the latter are touted as "renewable" or "environmentally safe," the process of manufacturing and maintaining them, and the industrialized lifestyles they help to perpetuate, are anything but. A sufficiently thorough accounting is likely to show that they are little more than an attempt to redistribute wind and solar energy using the rapidly depleting nonrenewable natural resources that go into their manufacture. A more beneficial, less harmful nuclear anti-technology would result in greatly decreased electricity use.

Mandatory technologies

THE TYPICAL MODERN living arrangement in the economically developed, industrialized parts of the world makes certain technologies mandatory. They tie you to the technosphere with an umbilical chord; sever it and your life is automatically imperiled. Consequently, these are the ones that grant the technosphere almost total control over your life—the ones which have the greatest negative effect on your autonomy, self-sufficiency and freedom.

Before we discuss them, we need to draw an important distinction: the technosphere's manifestations are of two sorts: **artifacts** and **flows**. The artifacts are all sorts of manufactured items of varying durability, maintainability and service life. The flows are actual flows of water, electricity, natural gas, sewage, bits of digital information flowing through the ether or through cable, food, gasoline, diesel fuel, heating oil, disposable products and recurring services. Crucially, all of these flows require a counterflow of money.

The artifacts can be enslaving (a television set is a prime example of such enslavement), but they can also be liberating (a fishing rod, reel and lure can liberate you from having to buy fish). They can be very long-lived: a collection of well-made hand tools with wooden handles is often passed from father to son. Some tools, if kept clean, oiled and sharp, can give many centuries of service and be used to build shelter and furniture and to perform your own maintenance. They can be produced using artisanal methods from industrial salvage, which will remain plentiful for geologic periods of time, steel especially. They can be repurposed and improvised, regardless of the initial use for which they were designed. They can be exchanged in gift and barter. They can be customized and fine-tuned to perfectly suit your body, the local environment and the culture. Most importantly, you can choose whether or not to use one of them.

The flows are none of these things. They are short-lived, lasting only as long as the substance in question is flowing and money is flowing in the opposite direction. They cannot be produced using artisanal methods out of industrial salvage, but must be kept up-to-date at additional expense. They cannot be repurposed and they often cannot be improvised. They can sometimes be exchanged in gift or barter, but this does not eliminate the need to pay for them. They cannot be customized or fine-tuned to fit with your environment or culture because they are mass-produced on a national or international scale and based on national or international standards. And often you have no choice but to use them and to somehow keep coming up with the money to pay for them.

The flows are described as "the grid," often specifically meaning the power grid—which distributes electricity in the form of alternating current across large distances—but the term is also extended to describe other technosphere services such as the water and gas mains and cable television. It is further used to describe an alternative lifestyle, called "off-grid living," which is often touted as a form of personal virtue. But getting rid of the grid, in whatever form, does not necessarily eliminate dependence on certain flows. Specifically, those who choose to live off-grid tend to be very heavily reliant on a certain high-priced disposable product, the car, and on a particular flow, gasoline, as well as the certain-to-follow recurring services—vehicle and road maintenance. And what the car allows them to do is to tap into many other flows, from propane and spare parts to prescription drugs and sanitary products.

Without all of these they would be unable to survive off-grid for very long, and all of these require money.

But there is an even bigger problem with off-grid living: giving it a name puts it in the crosshairs of the technosphere, slated for destruction, and all sorts of authorities, from local to federal, swing into action and impose local ordinances and federal regulations to make off-grid living difficult or illegal. The off-grid enthusiasts then start political protest campaigns to try to reverse these decisions, not understanding that their exact purpose is to send a message, and that in attempting to battle them politically they are acting as useful idiots who amplify that message. The battle they are choosing to fight is a losing proposition: the individual versus the technosphere. But perhaps the explanation for this poor choice of strategy is that they don't understand their enemy. Luckily, they can learn about their enemy—by reading this very book. And they will find that, in wresting control over their lives away from the technosphere, discretion and stealth are the better parts of valor.

To one extent or another, the grid, interpreted expansively, is considered mandatory just about everywhere. Going off-grid is normally an option only in certain remote, rural places; in cities, towns and other built-up environments it is more or less out of the question. If you attempt to disconnect from the water mains or the electrical grid, local officials may force you out of your house and confiscate your children for keeping them in what they deem substandard conditions. Not that substandard conditions are the least bit important to these authorities in most other ways: they are perfectly content to have your children breathe smog, drink water laced with heavy metals and residue from the hydrofracturing process, eat food contaminated with toxic, carcinogenic pesticides and herbicides and watch television programs and play video games that destroy their minds. Typically, they are not the least bit concerned about the harm *they* cause or fail to prevent. They only get upset over the harm *you* might theoretically cause or fail to prevent—or rather, they feign being upset, in order to exercise ever more control over you.

Personal standards

ALL OF THESE flows are mandatory in another sense: whether they come from the grid or not, you cannot survive without water, food, some way of cooking it, some form of illumination, a way of communicating with people and some amount of routine and emergency medical care. And here another form of standards comes into play—not official standards imposed by the authorities but your **personal standards**: how often you feel you need to wash, how small and simple a diet you can live on, how dimly you can stand your lights to burn, how rarely you communicate, how locally you travel and how well you can avoid doctors and other specialists.

Recall that one of the technosphere's goals is to cultivate in you maximum possible dependency, because by so doing it is able to extract from you maximum possible obedience.

- Everyone is conditioned to believe that a daily hot shower, which in North America averages 17.2 gallons (65.1 liters), is absolutely essential, while it is 20 times the average amount of water needed for survival (3 liters per day).
- We are accustomed to indoor illumination equivalent to hundreds of candles, whereas just one candle is sufficient in order to mend a garment or read a book.
- Anything short of a flush toilet causes us to recoil in horror, while outhouses and chamberpots were considered perfectly normal for many centuries.
- We want fresh fruit and seasonal vegetables in wintertime, flown in from across the planet, and a varied diet instead of one limited to a few staples and a few preserves as treats.
- We have been conditioned to demand continuous internet access, even though we could very easily adjust to composing e-mails and reading electronic documents off-line and doing all of our internet-accessing in a batch a few times a week.

- A lot of us can't imagine life without a personal automobile, and even going without a family fleet of cars—one for each adult member of the family—is often perceived as an unacceptable sacrifice.
- Disposable sanitary products are a particularly volatile subject, and a shortage of toilet paper, disposable diapers, sanitary napkins and tampons can easily result in a mutiny. Yet we have lived without any of these things for millions of years, and billions of us live without them today.
- When we grow old and become sick, the last thing on our minds is to lie down, say good-bye to everyone and pass on. No, we demand the latest high-tech treatments to grant us a few more months or years of frail, sickly existence, even if this means leaving our children without an inheritance or saddling them with debts. And the last thing we want to admit is that it is quite normal for death to be rather painful, even if it is relatively peaceful. No, we want to die *comfortably!*

Asking you to compromise on any of these standards is asking a lot. But if you refuse to compromise on them, then you will be forced to compromise on what's far more important—autonomy, self-sufficiency and freedom in particular. You will remain the technosphere's obedient servant until it fails, and then you will be left with nothing. It's your choice, and it's a hard one, but there are some tricks to making this choice both gradual and inevitable through the careful selection of technologies. In the next chapter we will explore just how simple, eternal and *naturelike* these technologies can be. But before that, let us delve into what is really keeping us from exploring such options.

Powerful technologies—weak humans

ONE ALMOST UNIVERSAL consequence of our using almost any sort of technology is that while it makes us more powerful and more capable as long as we have access to it, we end up quite a lot weaker and less capable the moment we lose it. In this sense, most kinds of technology, especially advanced industrial technology, is a Faustian bargain: in accepting it, we trade immediate power for eventual powerlessness.

This trend started a very long time ago. For a perfectly good example, look at your hands. See those fragile little slivers of keratin at the tips of your fingers—your fingernails? Those were once claws—capable of digging in the dirt and ripping flesh from bone. But once our ancestors learned to sharpen rocks and started using these new tools for digging up edible roots, fighting off animals and all sorts of other uses, the evolutionary pressure to maintain the strength of our claws fell away, and we eventually ended up with our pathetic little fingernails, which are only useful for picking away at things and have to be kept short or they break.

Here is another example: have you ever tried going on a raw food diet? Unless you choose your foods very carefully, you are quite likely to suffer and end up underweight and anemic from many bouts of food poisoning. The reason is obvious: since our ancestors discovered how to use fire to cook food, thus sterilizing it and making it much easier to digest, our digestive tracts evolved (or devolved?) to specifically deal with cooked food. Now we depend on being able to cook our food and have to maintain cooking fires, even when they cause deforestation (followed by soil erosion, which impacts our ability to grow food), as well as air pollution and respiratory problems from breathing the smoke.

Fast-forwarding to the most recent 4,000 years, the discovery of writing had an adverse effect on our memory. Preliterate societies pass along their knowledge verbally—from short stories to entire epic poems—and keep it all in their heads. Some of the oldest stories we have were preserved not because they were written down, but specifically because they were *not* written down but passed down orally. Languages constantly evolve, and oral traditions evolve with them, but writing freezes them, replacing living traditions with literary museum pieces expressed in dead languages which, after just a few centuries, can only be understood by scholars. Even after the invention of writing, for a long time reading and writing were specialized skills practiced by scribes and priests, but with the more recent advent of widespread literacy the need to remember texts verbatim became limited to actors, ✳ because everyone else can now simply look things up. But now that public education schemes in many parts of the world are failing and illiteracy is again on the rise it would be foolish to hope for a reversion to a preliterate status quo ante. Instead, the progression from oral learning to written learning terminates in widespread ignorance once literacy is lost.

Until quite recently, literate people, in spite of their memory deficits, could manage to write using very simple tools: ink pots, quill and bamboo pens and, in the Orient, brushes. But imagine what would happen now if you were to hand a lawyer a goose quill, a pen knife and a pot of ink, and ask her to draw up a Last Will and Testament or a Purchase and Sale Agreement. Nothing useful would come of that effort, I would imagine. Most people now are only capable of appearing professional if they have access to word processors, spell checkers and laser printers. Their spelling and their penmanship are too poor for them to be able to draft a convincing document by hand. We are in the process of producing a population that can only be considered literate as long as there are buttons for it to push; after that, all it will remain capable of is childish scribbles.

Childish scribbles may be sufficient to get by within a greatly simplified social realm, provided we are hardy. But we are not hardy either. We are born with brown adipose tissue—vascularized fat—which can directly generate heat without us having to shiver. But if we grow up in a temperature-controlled environment we lose it and thereafter depend on warm clothes and heated houses, or we wear ourselves out shivering and die of hypothermia. We have the ability to live with lots of minor infections and quite a parasite load, but indoor living, good sanitation and the constant use of hand sanitizers and antibacterial soaps deprive our immune systems of the inputs necessary to exercise them and maintain them at full strength. This results in a high incidence of allergies, asthma and autoimmune diseases, and should we ever find ourselves deprived of such hyper-sanitary conditions, our very survival will be at risk.

Similar examples can be served up endlessly. We can't undo a few million years' worth of hominid evolution even if we wanted to. Nor is it strictly necessary; we have ways to adapt. Deprived of modern comforts, many will die, but the strong and the hardy will survive. The industrial age has produced enough steel scrap to provide survivors with tool-making material for thousands of years. We even have some new tricks: solar cookers, efficient rocket stoves, compost digesters that produce methane for cooking, biochar kilns and so on. The quill pen, the inkpot and the letterpress can be brought back.

In the meantime, what we have is *the stopping problem*: when should we say that enough is enough? At what point should we say that we have been sufficiently pampered and rendered delicate by modern technology and start pushing back? Shouldn't we at least start resisting the effort to pamper and weaken us even further?

We also have the problem of *redefining success*: does figuring out a way to get paid for spending your entire life in front of an electronic screen of some sort sound like success to you? If so, it seems dangerously short-sighted, because that screen is going to flicker out, and then how successful will you be? Perhaps a better definition of success involves figuring out a way to survive and be happy *without* that flickering screen? In the meantime you might still use it for specific things. For instance, I am quite good with quills, but my publisher very reasonably requires the manuscript for this book to be submitted in electronic form. Also, I type at 90 words a minute but can only write legible longhand at 20 words a minute or so (even slower with a quill pen because of all the dipping). And so, I am banging out this book on a laptop. But in a permanent blackout situation I would look for geese and collect a few of their molted flight feathers to fashion into quill pens, then cook up some India ink using a 3,000-year-old recipe.

If we fail to address the stopping problem and continue to define success as success within the technosphere, then we will

continue to devolve in the direction of abject technological dependence until we become soft-bodied blobs, unable to support our *(happening now)*
own weight and having to spend our lives floating in vats of spe-
cially formulated liquids, unable to chew our own food but fed
through a tube, unable to move our limbs or use our senses but fed
information through neural implants. But I doubt that we will get
that far. As it is, many of us are so dependent on the technosphere
that an extended electrical blackout is likely, in numerous places, to
be accompanied by the stench of death—which is not the smell of
success by anyone's definition.

Unlimited harm potential

THE HARM/BENEFIT ANALYSIS is a useful tool for evaluating
existing technologies where both the harm and the benefits are
already known. It is less useful when examining new technologies that are yet to be put into widespread use, because we don't
yet know how useful they will be and, more importantly, we don't
know all of the unintended consequences their introduction will
bring on.

One usual consequence, which I hesitate to say is unintended,
is to make us weaker and more dependent on technology that
is beyond our control—in all of the many ways described in the
previous section and more. It is the self-interested intent of the
technosphere, as a sentient entity, to make us as dependent on it
as possible. And it is an easy result for it to bring about, because
we are quite easily seduced—because of our greed, love of comfort
and lust for power. The countervailing tendency, which is based on
a very healthy instinct, is to want autonomy, self-sufficiency and
freedom—but it is becoming a rarity. It is also usual for the introduction of new technology to result in unintended consequences
in the form of problems that can only be solved by introducing
newer, even more complex, advanced and expensive technologies,
resulting in yet more unintended consequences and so on until
there is no money left and no more solutions to be found.

Nuclear power industry

The nuclear power industry has built nuclear power plants having no idea what it would take to decommission them and permanently dispose of the spent fuel. And now it turns out that there isn't the money required to do the job. To use a specific example, a recent European Commission report found that the EU has just €150.1 billion of assets that are available of the estimated €268.3 billion needed to do the dismantling work.

Throw in just a handful of nuclear meltdowns, which, as the Three Mile Island partial meltdown in Pennsylvania in 1979, the Chernobyl nuclear disaster in the Ukraine in 1986 and the Fukushima Daiichi disaster in Japan in 2011 have shown, are very much within the realm of possibility, and the costs would multiply to a point where no national budget would be sufficient to control the damage. While the melted-down reactor at Chernobyl was more or less successfully encased in a sarcophagus and is currently not spreading radioactive contamination, the four meltdowns at Fukushima are known to be hemorrhaging a prodigious amount of contaminated water into the Pacific Ocean, and there is apparently no existing technology that can be brought to bear to stanch its flow—a perfect example of technology's unintended consequences requiring more, newer technology for their mitigation—technology that may never come into existence or may prove to be entirely unaffordable.

The worst-case scenario for nuclear technology is that social instability causes the entire nuclear stockpile to be abandoned to the elements, the structures within which it is contained breaking down over time and radioactive contamination diffusing into the environment and producing cancer rates so high that no human can survive to adulthood. This is the unlimited harm potential of nuclear technology.

Genetic engineering

Another technology with unlimited harm potential is genetic engineering, which is the artificial splicing of genes between organisms in ways that natural selection would never produce and the insertion into living organisms of completely synthetic, artificial genes that could never have been the result of natural evolutionary processes. A handful of transnational companies have generated genetically modified organisms (GMO) such as corn, rice, sugar beet and soy plants; most of the corn (maize) and most of the soy grown in the United States are now genetically modified.

One problem with GMO foods is that they are a slow poison. Laboratory animals fed a diet of GMO foods do not thrive and in as few as three generations lose their ability to reproduce. This information has been suppressed in the US and in the other countries dominated by Washington, because in Washington the handful of companies that are responsible for promoting and enriching themselves from GMO foodstuffs—such as Monsanto and Cargill—have excelled at something called *regulatory capture*: through their assiduous use of lobbying and political campaign contributions, these companies have ensured that the government agencies and the expert bodies charged with regulating them are staffed with their own operatives. So far, the only country that has been able to stand up to the American GMO mafia is Russia. It banned the import of all foodstuffs that could not be guaranteed free of GMO contamination. According to the strategic risk consultant F. William Engdahl:

> As a long-term two year independent laboratory rat experiment has demonstrated, a diet of GMO soybeans or GMO corn over a period of more than six months produces virulent tumors in the GMO-fed rats and excessive early mortality…. Since near 93% of USA corn today and 94% of its soybeans are GMO today, a safe rule-of-thumb is the precautionary principle–ban it unless proven GMO-free, which is precisely what Russian authorities have done. The Precautionary Principle is simply that, if regulatory authorities are not 100% certain it is GMO-free, prohibit it.[4]

---

4   "Russia Bans US GMO Imports," accessed June 18, 2016, http://journal-neo. org/2016/02/26/russia-bans-us-gmo-imports. The study cited in this article has received tremendous pushback from GMO advocates.

The Russian authorities have acted wisely, but as long as the US and the countries it dominates continue to grow GMO crops the danger persists even to the Russians. First, pollen from GMO crops can easily drift on the wind across the border and pollinate crops in nearby fields that are planted with natural, non-GMO crops, contaminating them. Because GMO crops produce nonviable seeds, when their pollen contaminates non-GMO varieties, it likewise makes them nonviable. The resulting loss of heirloom varieties, which, in their diversity, are far more able to withstand temperature extremes, droughts, floods, insects and nematodes than the one-size-fits-all GMO hybrids, poses a major threat to food security. As climate change and extreme weather place ever-increasing stress on crops around the world, the unintended consequence of genetic engineering will be not poor harvests but no harvests at all, resulting in famines and refugee crises around the world. Keep in mind that genetic engineering is altogether unnecessary—all being done for the sake for temporarily improved crop yields and, of course, corporate profits.

Second, as long as these countries are able to sell corn or soy-based animal feed made from GMO crops, GMO contamination can spread through animal products. Thus, the ultimate solution is rather radical. It involves an outright ban on food imports from all countries that have failed to sign on to that same outright GMO ban. There are now provisions in Russian law for regulating the safety of agricultural imports, but in practice the outright ban on GMO-contaminated products is turning out to be difficult and expensive to enforce.

Lastly, while it is hard but possible to protect Russian territory from GMO contamination, it is even harder to protect the Russian population. Every year some 30 million Russians travel abroad for business and pleasure to GMO-contaminated countries. Many of them report experiencing significant, inexplicable weight gain, especially during their visits to the US. Looking around in the US, it is obvious that they are not alone in this: three in four American men are considered obese, as are some 20 percent of children. So bad is the problem that names for new body parts had to be found: the "muffintop" (women spilling out of their pants); "moobs" (male breasts)...

There are many reasons for the American obesity pandemic, but one of the obvious culprits is glyphosphate—the active ingredient in Monsanto's herbicide Roundup™. It is a herbicide, but it also kills bacteria, including the beneficial bacteria in your gut which allow you to properly digest your food. Trace amounts of glyphosphate are now found in the blood of most Americans. The result is a population that is unable to properly digest its food, making it at once ravenous, anemic and obese.

It may be possible for Russia, which is a major agricultural producer and exporter, to stop GMO contamination at the border and, with the spread of Russian GMO-free agricultural exports, even to protect its citizens as they travel the globe, but this is not going to be so easy for many other countries which are unable to become self-sufficient in food and are politically unable to stand up to the "regulatory captives" ensconced in Washington. But it can be argued that even severe short-term economic damage from banning food imports from GMO-contaminated countries is better than the status quo, because the unlimited harm potential of genetic modification is a population that over a few generations becomes sickly, sterile and goes extinct.

Nanotechnology

Yet another technology with unlimited harm potential is nanotechnology, which is a set of methods for structuring materials at the nanometer scale. Hundreds of different applications of nanotechnology are already in commercial use, with numerous applications in automotive engineering, food packaging, medicine and other areas. Nanoparticles and nanomaterials have desirable technical properties, allowing engineers to design materials that are lighter, stronger, or have specific optical or chemical properties. The problem with nanotechnology has to do with what happens when it escapes into the environment. Very little thought has gone into determining the entire set of disruptions which nanomaterials can cause to living things, but all the indications are that they can be quite severe. Especially risky are the particles that are insoluble in water. When inhaled or ingested, they pass through cell membranes and can even penetrate the brains of animals. As they travel through the various food chains, they can do untold damage by causing certain key species to go extinct.

Of all the technologies with unlimited harm potential, nanotechnology is the worst. Countries have the ability to disallow the use of nuclear technology within their borders. They can also ban the import of GMO-contaminated seeds and foods. But it is impossible to ban nanotechnology, because it is part of the proprietary know-how of hundreds of separate manufacturing companies from soft-drink bottlers to information technology companies. There are no tests to determine whether a manufactured item contains nanotechnology. And while the harm from radioactive contamination and GM-contaminated foods can be evaluated within the laboratory, it is simply impossible to assess the full extent of the damage that nanoparticles can cause to the biosphere.

The harm/benefit hierarchy

TECHNOLOGY, IN AND of itself, is neither good nor bad, and it is essential for survival. Whether it helps us or harms us is a question of whether and how we use it. Our job is to pick and choose carefully, to embrace technologies that liberate and empower us and to look for ways to avoid or eliminate the ones that weaken us, make us dependent on outside interests and forces and can even result in our extinction as a species (yes, there are a few such technologies).

If we do this job well, then we will shrink the technosphere. The mechanism by which the technosphere will shrink is a simple one to describe but rather difficult to evaluate in a quantifiable fashion. But sometimes such quantification isn't necessary; all that's needed is the ability to predict with confidence that the required result will be obtained through some finite, physically possible amount of effort. For example, there is no need for precise mathematical models to estimate how many whacks with a stick it will take to break open a piñata; it is sufficient to know that some reasonable number of whacks is enough to get at the candy inside it. And so it is here: some amount of effort, which we humans, being adaptable and resourceful, are most certainly capable of making if we are sufficiently incentivized and motivated, will produce the result we are looking for.

The technosphere expands when it gains *efficiency*. The efficiency in question is not some relative measure of the amount of useful output for a certain amount of input. Rather, it is the systemic efficiency of the technosphere as a whole in controlling us: do its actions, all other things being equal, give it a greater measure of control over us? If we select technologies that cause us to relinquish control to the technosphere, then we are doing its work for it, making it more efficient. Conversely, if we select technologies that specifically deprive the technosphere of means of control or make the exercise of control more expensive in terms of time, resources and energy, then we reduce its efficiency and its scope.

This effect is automatic. The technosphere's emergent intelligence is the intelligence of a machine. Its internal programming is such that it *always* acts rationally in its own internal self-interest. For the technosphere, the ends *always* justify the means to the exclusion of every other consideration. These ends are limitless growth and expansion, complete domination of the biosphere and complete control over us humans. If we succeed in thwarting it to a point where it reaches diminishing, then negative returns, and increased effort leads to *decreased* results, then, being rational, it will have no choice but to decrease effort... and shrink. This process, if taken far enough, will reduce it to a set of mere instrumentalities from which we can choose the ones we like à la carte—no longer able to pursue its own agenda but pressed into service if needed and allowed to languish and disappear if not.

Cost-benefit analysis

Please note that harm/benefit analysis has *nothing* to do with its commonplace cousin, cost/benefit analysis. The technosphere *loves* cost/benefit analysis, because its underlying assumption—that everything can be quantified in terms of monetary value—makes everything and everyone into a commodity, providing an objective, rational basis for expanding the technosphere's control over us and the biosphere. The idea of cost presupposes a pay-off: there is no harm as long as someone pays, even if it is blood money. The idea of harm presupposes something else: once it is done, it cannot be undone, any more than a rape can be undone by buying the victim some flowers. Those who have caused harm can bemoan their fate while awaiting the consequences.

And so, please keep in mind that harm is *not* a cost, and estimating harm in terms of its cost would be a completely wrongheaded approach. Neither the biosphere nor the human spirit can be appeased with an apology; nor will they accept any sort of monetary restitution for the harm caused to them. You, on the other hand, may be bought off. But asking how much your autonomy, self-sufficiency and freedom are worth in financial terms is equivalent to asking how much you are worth as a slave. With regard to the harm that the technosphere inflicts on the biosphere, we should seek not compensation but curtailment of its activities. With regard to slavery, we should strive to not be slaves and, failing that, work on becoming the most worthless slaves we can be— if possible, worse than worthless, causing harm to our masters in ways that are difficult to detect.

Technologies that should be disallowed

Applying the harm/benefit analysis to comparable technologies makes it possible to determine which technologies are preferable, and to choose a reasonable cut-off point. Above that point are the technologies that are either too harmful or not sufficiently beneficial to deserve a place in your technology toolkit.

At the top of the hierarchy are the technologies that have unlimited harm potential: nuclear technology, genetic engineering and nanotechnology. Here, the harm is potentially infinite, while the benefits are finite, if indeed they exist at all once all of their costs are taken into account. For nuclear, the benefit is increased electricity generation but at the cost of massive government subsidies, without which no nuclear power industry would ever have been possible. The benefits of nuclear power disappear completely if one includes the costs of dismantling nuclear installations and of safely sequestering radioactive waste over geologic periods of time. For genetic engineering, the benefit is temporarily increased crop yields while the harm lies in the destruction of food security and the potential sterility of the human race in just a few generations. For nanotechnology, the benefit is in the better performance of engineered materials and the harm is, at this point, impossible to predict, but it is potentially fatal to all higher life forms. To determine the position of these three technologies within the harm/benefit hierarchy, we take their infinite harm potential and divide it by their finite total benefit B:

$$HBR = \frac{\infty}{B} = \infty$$

Therefore, they go at the very top of the harm/benefit hierarchy, above any reasonable cut-off. Consequently, these three technologies should not be used at all, under any circumstances. Another consequence of this calculation is that any potential anti-technologies that can target and neutralize technologies with unlimited harm potential are by far the most beneficial of all anti-technologies and therefore deserve very careful consideration. For example, one might use genetic engineering techniques to produce particularly hardy and invasive herbicide-resistant weeds and release them into the wild. They would make herbicides ineffective, bankrupt Monsanto and derail the entire GMO train. They would cause limited harm in place of the unlimited harm caused by GMOs.

Since many weeds have already evolved herbicide resistance,[5] one would, in theory, only be helping nature. Such approaches to anti-technology are not without their dangers: in a "sorcerer's apprentice" scenario, they could create anti-technologies more harmful than the technologies they attempt to neutralize. But they also offer great opportunities because the most potent way of disrupting a complex technology is through internal sabotage.

Technologies that may be allowed

Lower down in the hierarchy are the technologies that cause finite harm and provide finite benefits. Here, the 32 aspects listed in the table on page 100 allow the harm/benefit ratio to be determined with some accuracy.

Some of these aspects overlap in a variety of ways. For instance, something that is used frequently by an entire community would be best if it were commercial-grade rather than consumer-grade because otherwise it would wear out quickly and not be maintainable. Thus, it is reasonable to rule out the individual use of many technologies—such as washing machines, automobiles and computers—but allow their use by the entire local community, by providing a community laundry, a fleet of cars and trucks, and a community library equipped with computers.

The new/(re)used aspect is particularly important: if something already exists, then it doesn't need to be manufactured and therefore causes less harm, because all manufacturing processes deplete nonrenewable natural resources and pollute the environment. Used equipment tends to be cheaper. It is often in disrepair, in which case it can only be used once it has been repaired. But if it is repaired, then that proves that it *can* be repaired and that the skills needed to maintain it exist within the community, helping the community become more self-sufficient.

---

5   Veronique Dupont, "US 'superweeds' epidemic shines spotlight on GMOs." http://phys.org/news/2014-01-superweeds-epidemic-spotlight-gmos.html.

Let us try working through a specific example: choosing a pas-
senger car to be used as part of a community fleet. (We will assume
that in this community private cars are considered above the cut-
off in the harm/benefit hierarchy.) For the sake of the argument,
let's consider two rather extreme choices: a 1976 Jeep Cherokee
and a 2016 Tesla S electric vehicle.

- The Jeep, and most of the parts needed to keep it running, come
  from the local junkyard. They don't need to be manufactured, causing no damage. The Tesla comes from a modern, high-tech factory
  with a huge environmental footprint, and the process of manufacturing it depletes nonrenewable natural resources (lithium
  especially) and causes environmental damage.
- If the Jeep needs "new" parts, these can be remanufactured by any
  number of local machine shops using parts obtained from any
  number of local junkyards. The parts for the Tesla can be sourced
  from just one or two giant factories. Most of the components come
  from overseas. A single tsunami, earthquake or flood can disrupt
  the supply for months.
- The Jeep is practically free; the Tesla sets the community back at
  least $70,000.
- The Jeep probably arrives pre-broken in a variety of ways, and if it
  can be fixed, then some member of the community knows how to
  keep fixing it. But if anything goes wrong with the Tesla, only the
  Tesla dealership can fix it, using proprietary tools.
- Driving the Jeep causes environmental damage because it emits
  pollutants and greenhouse gases, while the Tesla is supposedly
  zero-emission—but this advantage is negated once we take into
  account the harm caused by connecting it to the electrical grid.
  Even if the electricity comes from renewable sources, such as
  wind and solar, the solar panels and wind turbines have to be manufactured too, and that manufacturing process is certainly not
  zero-emission.
- Should the electric grid fail, the Tesla becomes useless while the
  Jeep keeps running and can even be used to charge batteries for

other uses. Even if gasoline becomes unavailable, the Jeep can be converted to run on methane from a digester because it is old enough to have an old-fashioned carburetor rather than the more "efficient" fuel injectors used in newer engines.

- The Tesla uses lithium batteries which wear out; lithium is a nonrenewable natural resource that is forecast to become scarce within a decade or so. This will render the Tesla useless because no replacement batteries will be available for it. The Jeep uses a lead-acid battery, which can be brought back to life once or twice by draining it, flushing it out with baking soda and water and then filling it with fresh acid (which is a cheap chemical waste product). Even after the lead plates wear out, it can be remanufactured by making new lead plates by hand.

- The Tesla has just one use—transporting passengers over roads— while the Jeep can also haul loads and tow trailers over many kinds of terrain.

- Finally, there is also the following consideration, which makes the choice of Tesla seem altogether ridiculous. The Tesla is supposedly better in some way because it doesn't burn oil. But it is designed to drive on a sea of oil in the form of asphalt and doesn't do well on dirt roads. Asphalt is a waste product generated by refineries in the process of making gasoline, diesel fuel and other petrochemicals. Thus, the fact that the Tesla does not directly burn petrochemicals is entirely beside the point. Nobody will ever run a petroleum refinery unless there is demand for gasoline and diesel. Unless the refineries are operating, no asphalt will be produced, the roads won't get paved, and the Tesla would, once again, become useless. The Jeep is specifically designed to do well on rutted, potholed dirt roads.

A similar analysis can be applied to just about any piece of technology that causes some amount of harm and provides some benefits, and it will produce comparable results: something that is simple, reused, is easy to maintain and is useful to the community in multiple ways will always end up lower on the harm/benefit scale than something that is complex, newly manufactured, requires expert maintenance and is only useful to an individual.

Zero-harm technologies

Finally, at the bottom of the harm/benefit hierarchy are all the technologies that cause no harm at all. Since their harm is zero, they don't have a harm/benefit ratio:

$$HBR = \frac{0}{B} = 0$$

Instead, they should be prioritized just by looking at the benefits they bring, with the most beneficial technologies seeing the most use.

The best of these technologies can be called *naturelike*: they are human adaptations of things nature has produced as evolved traits in other species. The next chapter is devoted to examining naturelike technologies.

The dangers of nonexistent technology

WHILE MANY TECHNOLOGIES cause harm by their existence, others cause harm by their absence. A good example of such missing but vital technology is on display in the ongoing disaster at the Fukushima Daiichi nuclear power station. There, not one, not two but three missing technologies are causing major problems:

I. There is no way to find out where the nuclear fuel went. It melted through the bottom of the containment vessels in three of the four reactors, but that's as much as anyone can tell you. The area is far too radioactive for technicians to work in. After many failed efforts, it has been conceded that there is no way to build robots that are sufficiently radiation-hardened to survive in this environment either. After much effort, some robots were sent in, beamed back a few grainy black and white images and died without discovering any trace of the melted-down fuel.

2. What remains of the melted-down reactor vessels is being cooled by pumping in water. If it weren't for this effort, the melted-down fuel would burst into flames and emit a pall of radioactive contamination. In the process, the water becomes severely contaminated with radioactive isotopes, some very long-lived, and has to be stored in tanks, which are being hastily welded together. There is no technology to effectively decontaminate this water quickly enough, and the tanks are proliferating around the site. In time, they will corrode and start to leak.

3. Groundwater infiltrates the site and then leaks out to the Pacific Ocean, carrying radioactive contamination with it. An effort is being made to construct an ice dam by freezing the ground between the reactors and the ocean, but so far the results are far from positive. There is no technology to isolate the groundwater around the reactor sites from the surrounding groundwater.

This is a particularly stark example of the dangers of missing technology, but there are numerous others, and we should not simplistically assume that minimizing the use of technology is always an unambiguously good thing. Instead, we should always be on the lookout for technologies that can cause great harm by their absence. In the case of technologies having to do with dismantling nuclear reactors (even ones that have melted down) and long-term sequestration of spent nuclear fuel, their absence can even be said to have unlimited harm potential.

Relative harm

A PARTICULARLY USEFUL saying to keep in mind is "Better is the enemy of good enough." If we already know that our current lifestyles destroy the biosphere, cause harm to us and to society and empower the technosphere, but we cannot conceive of a perfect alternative scheme, then we should try to think of an imperfect, perhaps downright primitive, perhaps barely workable scheme that would nevertheless be better. To this end, the harm/benefit analysis can be applied to your lifestyle one element at a time and used to look for opportunities to make specific changes that may provide somewhat fewer benefits but cause significantly less harm. This will allow you to ratchet down the harm/benefit hierarchy one notch at a time.

This, then, is the way to get started on a path down the harm/benefit hierarchy, with the goal of eventually reaching the level of mostly naturelike technologies that cause no harm at all. Whenever an element of technology—one that you (at least at the moment) cannot do without—is in need of replacement and several alternatives present themselves, calculate the harm/benefit ratio for each one of them and pick the one with the lowest ratio.

If an element of technology is particularly harmful, yet indispensable, consider how it is used rather than what it is in and of itself. For instance, if you are, for the time being, stuck in a location where it is impossible to survive without a car, consider not the car itself but how it is being used. Can it be used as a fleet vehicle, shared by several people? Is it possible to combine trips with other people? Is it possible to replace it with a car that can be serviced and maintained by someone within your community on a barter basis instead of having to pay an outside mechanic? Most importantly, can it be driven less?

If practiced over a period of time, this technique will lead to lessened dependence on the technosphere and, if it is practiced by enough people, it will shrink the technosphere. It may seem minor in the grand scheme of things, but it is something that you can actually do. Just as, if you started feeling out of shape and decided that you need some exercise, you wouldn't start by bench-pressing 300 pounds or running a marathon, so here you start small and work your way up. This technique is perfectly rational and perfectly legal and will probably save you money, time and aggravation. It is what you should try first.

5
=

NATURELIKE TECHNOLOGIES

The germ of an idea

INSPIRATION CAN SOMETIMES arrive from odd places. That's how it happened with the idea for this book: it started with one word—*naturelike*—which I got, of all places, from listening to a speech by Vladimir Putin, Russia's president. On September 28, 2015, while addressing the UN General Assembly, Putin proposed "implementing naturelike technologies, which will make it possible to restore the balance between the biosphere and the technosphere." It is necessary to do so to combat catastrophic global climate change, because, according to Putin, CO_2 emissions cuts, even if implemented successfully, would be a mere postponement rather than a solution.

I hadn't heard the phrase "implementing naturelike technologies" before, so I Googled it and Yandexed[1] it, and found that all the references were to that same UN speech. Apparently, it was Putin who coined the term *naturelike*. Now, Putin has received a remarkable amount of negative press in the West, which may have affected your impression of him. The negative press mostly has to do with the oligarchs who own Western media companies

---

[1]   Yandex is an internet search engine that is popular in Russia.

and the fact that they are unhappy with Putin's policy of putting the interests of the Russian people first and those of the oligarchs a distant second.

But that is all quite beside the point. For the sake of this discussion, all you need to understand about him is this: this man does not throw words to the wind. As with the other phrases he's coined, such as *sovereign democracy* and *dictatorship of the law*, it was a signal that the rules of the game had been changed. In each of these two cases, the newly coined phrase served as the cornerstone of a new philosophy of governance, complete with a new set of policies. And, again, regardless of your opinion of him, his 80-percent-plus approval rating among Russians should suffice to convince you of just one thing: his policies tend to be quite effective.

In the case of *sovereign democracy*, it meant methodically excluding all foreign influences on Russia's political system, a process that culminated recently with Russia, in tandem with China, banning Western NGOs and those financed by the billionaire George Soros specifically. Previously, these organizations were making futile attempts to destabilize Russia and China politically. Other countries that find themselves in trouble with the Color Revolution Syndicate, which seeks to destabilize countries in order to make them act in conformance with the dictates of the neoconservative elites that are entrenched in Washington, DC, can now follow Russia's and China's best practices. We can see the effectiveness of this policy reflected in the utter desperation of these neoconservative elites, who, along with their Russian puppets, such as the former oil tycoon and convicted tax cheat Mikhail Khodorkovsky and the chess player-turned-politician Gary Kasparov, have attempted to manipulate and destabilize Russia. Thanks to Putin's sovereign democracy they have been rendered irrelevant, with their self-important pronouncements now amounting to little more than near-apoplectic ranting.

In the case of *dictatorship of the law*, it meant either explicitly granting amnesty to or legalizing, or explicitly outlawing and destroying, all types of illegal or only semi-legal social formations first by focusing on the criminal gangs and protection rackets that proliferated in Russia during the wild 1990s. It is now expanding into the international sphere, where Russia is now working in Syria, in cooperation with its regional partners, to destroy the products of illegal Western activities such as ISIS[2], along with other US-trained and US-armed terrorist groups that have been financed by Saudi Arabia and Turkey. *Dictatorship of the law* means that no one is above the law, not even the CIA or the Pentagon. As of the time I am writing, this policy has produced some limited yet remarkable successes in stabilizing Syria by purging it of foreign-funded terrorist groups. Note that Russian armed forces have been operating in Syria legitimately: they are in Syria by invitation from Syria's legitimate, elected government; all other foreign actors who have bombed or invaded Syrian territory are, technically, war criminals. Thus, Russia's successes in Syria are also victories for international law.

Since Putin seems to have an uncanny ability to make his words stick by altering reality to conform with them, it makes sense to carefully parse the phrase "implementing naturelike technologies" with the goal of gaining a better of understanding of what Putin meant by it, and what he might be up to. This particular phrase is harder to parse than the previous two, because the Russian original, внедрение природоподобных технологий, is laden with meanings that its English translation does not directly convey.

"Внедре́ние" (*vnedrénie*) can be translated in any number of ways: implementation; introduction; inoculation, implantation (of views, ideas); entrenchment (especially of culture); enacting; advent; launch; incorporation; adoption; inculcation, instillation; indoctrination. Translating it as "implementation" does not do it justice. It is derived from the word "не́дра" (*nédra*) which means "the nether regions" and is etymologically connected to the Old English word

---

2   ISIS is a terrorist organization that goes under several names, including ISIL, Daesh, Islamic State and Islamic Caliphate. It has been banned by the Russian Federation.

neдera through a common Indo-European root. In Russian, it can refer to all sorts of unfathomable depths, from the nether regions of the Earth (where coal, oil, gas and various ores and minerals are found) to the nether regions of human psyche, as in the phrase "недра подсознательного" (*nédra podsoznátel'nogo*, the nether-regions of the subconscious). It can very well mean "implantation" or "indoctrination."

The word "природоподобный" (*priródo-podóbnyi*) translates directly as "naturelike," although in Russian it has less of an overtone of accidental resemblance and more of a sense of active conformance or assimilation: "beseeming of nature." This word could previously only be found in a few techno-grandiose articles by Russian academics in which they promote vaporous initiatives for driving the development of nanotechnology or quantum microelectronics by simulating evolutionary processes or some such. The basic thrust of their proposals seems to be that even if our devices become too complex for human brains to design, we can let them design themselves, by letting them evolve like bacteria in a Petri dish. But it is hard to see how this interpretation of the word is at all relevant. Also, based on what Putin said next, we can be sure that this is not what he had in mind:

> We need qualitatively different approaches. The discussion should involve principally new, naturelike technologies, which do not injure the environment but exist in harmony with it and will allow us to restore the balance between the biosphere and the technosphere which mankind has disturbed.

These were the two sentences that made an alarm bell go off in my head. I had thought that same thought before, but I had never heard it expressed quite so cleanly and crisply and certainly not before the United Nations General Assembly. And so I thought, "OK, why don't I start working on that?"

But what did he mean by "technologies"? Did he merely mean that what we need is a new generation of eco-friendly gewgaws and gizmos that are slightly more energy-efficient than the current crop? Again, let's see what may have been lost in translation. In Russian, the word технологии (tekhnológii) does not directly imply industrial technology and can relate to any art or craft. Since it is obvious that industrial technology is not particularly naturelike, it stands to reason that he meant some other type of technology, and one type immediately leapt to my mind: **political technologies**. In Russian, this term is written as one word, политтехнологии (polit-tekhnológii), and it is a word that sees a lot of use in Russian public life. At its best, it is the art of rapidly shifting the common political and cultural mindset in some generally beneficial or productive direction. At its worst, it is an underhanded attempt to manipulate public opinion for private benefit.

Putin is a consummate political technologist. His current domestic approval rating stands at 89 percent—the remaining 11 percent disapprove of him because they wish him to take a more hard-line stance against Western aggression. It makes sense, therefore, to examine his proposal from the point of view of political technology, discarding the notion that what he meant by "technology" is some sort of new, slightly more eco-friendly industrial plant and equipment. If his initiative succeeds in making 89 percent of the world's population speak out in favor of rapidly adopting naturelike, ecosystem-compatible lifestyles, while the remaining 11 percent rise up in opposition because they believe that the rate of their adoption isn't fast enough, then perhaps climate catastrophe will be averted or, at least, its worst-case scenario—the one that includes near-term human extinction. I hope you will agree that, given the scarcity of other such proposals from supposed world leaders and given the success of his previous initiatives, this new one might be worth a try.

Before we proceed any further in describing how political technologies can be used to bring about the sort of dramatic social change that might grant humanity a new lease on life, let's ask what "naturelike technologies" might be like. By "naturelike" we mean something that is in balance with nature (its rhythms, both diurnal and annual, and its cycles: of water, carbon dioxide, organic and inorganic nutrients) and with the uninterrupted flow of human generations, which preserves local languages and cultures, along with their intimate knowledge of complex, diverse natural environments. By "technologies" we mean the practical know-how, passed from generation to generation, which one needs in order to survive—not any fancy gadgets or machinery, not the "internet of things," nano-this or genetically-engineered-that.

Of course, there must also exist political technologies that can sustain and defend such an effort, especially against the predations of profit-driven psychopaths who have imperiled human survival through rapid resource depletion and out-of-control industrial development. One of the chapters that follow is devoted to political technologies, explaining what they are and how they are being used for both good and evil, and sketching an outline of how they can be used to bring about the needed change.

Village life

"LIFE IN HARMONY with nature" has the ring of a hackneyed phrase: everybody talks about it but nobody does anything about it. So far, the discussion of what *naturelike* means has been rather theoretical. But to me it is not an abstraction: I have seen it for myself, and I have lived it. Let us now shift gears and dive straight into the topic. I will focus on what I know; the world is big, and it is impossible to say anything specific about it without talking about a specific piece of it. And so I will talk about the northern half of the Eurasian landmass. But I hope that my observations and extrapolations can be extended to other habitats around the world—at least the ones that will remain survivable as the planet warms, the seas rise and climate chaos reigns.

· · ·

WHILE I WAS growing up in the USSR in the late 60s and early 70s, every summer, from age five to age nine or so, my family would take off in some direction, east, west or south, on trips that could, in some respects, be described as trips back in time. We spent one summer in a village so out-of-the-way that the locals demanded to know how we ever found out about it. We hadn't known about it, neither did the authorities in the regional center, and the locals seemed keen to keep it that way.

We simply tagged along with a geological survey team that was doing seismic testing, blasting its way along a hydrocarbon seam. Our method of transportation was the *smótka*—a reel truck that bumped along rutted dirt roads running cable between sensors stuck in the ground, triggering small explosive charges and recording the resulting seismic data as jagged lines on spools of graph paper spewed out by a seismograph inside the truck. We simply stumbled across this village during our wanderings and decided to stay for the summer and catch the *smótka* again for the trip back to civilization. The locals were happy to provide us with accommodation by opening up a disused log cabin, laying down floorboards and providing us with a starter kit of sorts for making it livable.

It was a very poor village, with half of the intact log cabins boarded up, its few remaining residents in rough shape. At night wolves and bears roamed the village, and meat, when we managed to get some from passing truck drivers, had to be buried outside in a pit with boulders piled on top. As the summer wore on and the wolves and bears got better at digging, the pit became deeper and the boulders larger.

We spent another summer in a village in the vicinity of Rybinsk, near lands flooded by the construction of hydropower dams, where most of the transportation was by boat over flooded land and where the older people spoke an impenetrable Finno-Ugric dialect whose name they didn't know. This village was more prosperous; many households had cows, and their hooves churned up the dirt road that ran through the village center into a knee-deep wallow whenever it rained. I remember getting my rubber boots stuck in the mud and trying desperately to pull them out before the oncoming herd of cattle got to me, which included some quite ornery bulls that would periodically decide to chase me.

There were other summers too—with a family in Estonia on a homestead that had barely changed since the Middle Ages and had become something of a museum, with an impressive collection of wrought-iron candelabras that had actual candles in them and were actually used for illumination, there being no electricity. Another summer was spent on another homestead in the Transcarpathia region of western Ukraine—a working farm, where I got a chance to herd cows on horseback (which I liked), helped with making hay (which I hate to this day) and slept in a hayloft with the family's four rather uninhibited daughters (which I liked a lot).

We spent two summers in a hunting lodge on Lake Ladoga[3] in Karelia which had once belonged to Baron Gustaf Mannerheim. His days of owning it had only lasted during the short period when Karelia was no longer part of the Russian Empire's Grand Duchy of Finland but had not yet become the Karelo-Finnish Soviet Socialist Republic. By the time we got to the lodge it had been nationalized, turned into a resort and given over to the Soviet Composers Union, of which my father was a member.

There, I became obsessed with fishing, trolling from a rowboat for fresh-water pike which hid in the deep crevasses that ran along the bottoms of the fjords. The angle of the sunlight had to be just right in order for the pike to see my lure, which I made to dance in an apparently convincing impersonation of a dying fish. We ate the pike after hot-smoking it in a metal box filled with alder twigs—a process that made this normally very tough and bony fish tender and delicious.

Wherever we happened to be spending the summer, most of our days were spent wandering around the woods, foraging for berries, mushrooms and whatever else the woods had to offer.

---

3   Lake Ladoga is the largest lake in Europe. Its surface area is 17,700 km^2 (6,800 sq. mi).

Everywhere, the woods were a maze of trails worn by the movements of both animals and people over many thousands of years.

Wilderness as a state of mind

UNLIKE WHAT AMERICANS like to call "wilderness" or, worse yet,
"unimproved land," the land I had roamed in my childhood was
land that couldn't possibly be improved. It was already perfect—
alive and full of spirit, where animal and human souls commingled
over thousands of years of uninterrupted harmony. In comparison,
the North American landscape, with its national parks, marked
hiking trails headed by large parking lots and the rest of it posted
"no trespassing," is a dead landscape, bereft of spirit and meaning,
maintained as a wilderness only because wilderness is thought
to have uses such as recreation and conservation. That landscape
is artificial: a mental construct overlaid on a natural realm that is
considered alien. To an American, the map is the landscape; to a
Russian living deep in the countryside, a map is evidence that you
might be a government official or, worse yet, a foreign spy.

In most places in Russia you can walk in any direction for
almost any distance, guided by memory and instinct rather than
trail markings or a map. Rather than follow a marked trail sin-
gle-file as Americans tend to do, in what to the untrained eye looks
like some sort of prison walk, in Russia people fan out across the
landscape and keep in touch by yoo-hooing back and forth. Even
young kids tend to wander the woods on their own, because Amer-
ican-style safety-consciousness is nonexistent and would probably
be considered harmful—a way to raise nincompoops.

Bringing back the village

BUT THE GLORIOUS Russian woods of my youth, so full of wonder and spirit, were also quite empty of people. In many places, we
would happen across abandoned homesteads: trees growing from a
foundation pit, the house that had once stood over it long decayed

into a moss-covered mound; a stout willow sprouting from a caved-in dug well—nature swiftly reclaiming the land. About the only structure left standing would be the one, rather impressive, bit of masonry found in every traditional Russian village house: the Russian stove (of which more later). Sometimes we would stumble across the remains of an orchard and a garden, the fruit trees—apples, pears, plums—still bearing fruit, along with the bushes and cane—currant, raspberry, gooseberry—but the potato field and the vegetable garden would have already reverted to woodland.

This emptying of the rural landscape in Russia was one of the worst outcomes of the 20th century: collectivization and rapid industrialization following the Revolution of 1917 drove people out of the villages and into the cities. The centuries-old patterns of local democratic self-governance and self-reliance were destroyed in a single generation. Old family farms were replaced by large communal farms and centrally planned agricultural production schemes. These, as it turned out, were an unmitigated failure, forcing the USSR to resort to importing grain from the US and Canada[4] on credit, paving the way to its eventual destruction at the hands of its foreign creditors. Luckily, this failure was temporary; a quarter-century after the Soviet system fell apart, Russia is once again one the world's main agricultural producers and exporters, taking first, second or third place in the production of most agricultural commodities and is now poised to become the world's main exporter of organic food, uncontaminated by genetic modification.

Although there is quite a lot of mechanized industrial agriculture in Russia, for export commodities especially (Russia's Rosselmash stacks up well against John Deere) a lot of the food is still grown on small plots, which tend to be very productive, and their produce, sold through ubiquitous farmers' markets, is of

---

4   This development caused Winston Churchill, a notorious Russophobe, to quip in a letter: "I thought that I will die of old age, but now I know that I will die of laughter." Alas, the reason for his laughter was transitory; as of 2015, Russia is once again the world's largest grain exporter.

higher quality. Recent political developments have given local producers a big boost. Western sanctions illegally imposed on Russia following the putsch in Kiev in February of 2014, Crimea's referendum and the civil war in eastern Ukraine resulted in Russian countersanctions, which banned food imports from the offending Western nations. At the same time, lower oil prices and export revenues and an attack by Western speculators that drove down the exchange rate of the ruble made imports more expensive. Sanctions notwithstanding, a lot of the food imports are never coming back. The Russians have started paying close attention to where their food comes from, plus Russia has banned all genetically modified products, cutting off imports from the almost entirely GM-corn-fed Americans.

A number of new initiatives and new legislation are making it easier for people to go back to small-scale farming. Certain categories of people, such as veterans and young families with children, now receive free parcels of land from the government.[5] Income tax, which is normally a 13 percent flat tax, drops down to just 6 percent for those who take up farming. Other factors, such as the widespread penetration of cell phone service and internet access and the growing popularity of homeschooling (by Russian law, schools have to compensate homeschooling parents) are helping make rural living more popular. Reports that periodically surface in the social media from those who have moved out to villages tend to paint idyllic portraits of rural living. Slowly but surely, the landscape is being resettled.

This trend is in many ways a reversion to norm: over much of its millenium-plus-long history, Russia was a country of many small towns, numerous villages and countless isolated homesteads. This pattern of habitation suited the landscape, which is vast but provides rather diffuse resources to a preindustrial economy. Until quite recently most houses in Russia, even large ones, were built

---

5   This initiative is currently in a pilot stage, is limited to lands in the Far East and is only open to locals.

with wood, limiting their lifespan. Wood was, and still is, plentiful, but stone is quite scarce in many places, limited to some scattered boulders which many ice ages ago were rolled across the landscape by glaciers. Because of this, very little has remained in the way of ancient buildings or ruins: a few churches and a few forts.[6] Of course, this changed with industrialization and the advent of cement, brick and reinforced concrete, but for many centuries before that Russian civilization left scarcely any permanent mark on the landscape—hardly anything that nature could not quickly reclaim by fire or decay.

A good way to inhabit the landscape

THE HUMBLE, RUSTIC Russian village cabin, with its log walls and thatched or shingled roof, is physically impermanent: the logs rot; the thatch had to be replaced every few seasons. All that can be done to extend the life of the structure, which often simply sits on wet ground over a shallow foundation pit, is to periodically replace the bottommost logs, but even then, after a few decades, the entire structure has to be abandoned—disassembled and cut up for firewood, burned in place, or simply allowed to decay into a compost heap. But as an easily replicated technology, it is ageless: a perfect set of adaptations to a difficult and demanding environment that have been honed to perfection over many centuries. It, along with the lifestyle and the practices associated with it, is a very good example of an all-encompassing naturelike technology.

The house

The Russian village log cabin, called *izbá*, has many features that are perfectly adapted to its harsh northern environment. It is not the purpose of the book to describe all of them, but to give a sense of what is meant by *naturelike technology*, let us describe just two of them.

---

6   The forts are called "kremlins." Most ancient Russian towns have one.

The first of these is that the *izbá* is not entered directly but through an unheated space called *séni*. Rather than attempting to translate it, this term is better left transliterated because no combination of English terms—entry, hall, mud room, pantry, anteroom, airlock, workshop—is adequate to describe it. It is a completely enclosed, secure but unheated space that shares a wall with the main room, is often just as big as the main room and has a myriad uses. It is used to hang outdoor clothing, to store skis, snow shoes, fishing tackle and bait, tools, food and much else. Thanks to the *séni*, where almost everything is stored, the main living space is not filled with junk but can be left sparsely furnished, making even a small space seem spacious. In the main room, there are usually a couple of benches, wide enough to sleep on in the summer, and a table. There are also a few shelves for books, a wardrobe, a cupboard and a chest for valuables. The rest of the living space is taken up by the second feature of the *izbá* which we will discuss below: the Russian stove.

Compare this arrangement with the typical North American suburban or rural house. The many functions of *séni* are served by a number of architectural features: a porch, which is sometimes screened and full of junk; a pantry; a basement; and a garage which is normally too full of junk to leave room for a car. Likewise, the many functions of the living space in an *izbá* are served by a living room, a kitchen, perhaps a study or a family den and some number of bedrooms. The benches in an *izbá* replace kitchen chairs, office chairs, armchairs, sofas and beds; the table takes the place of kitchen counter, the dining table, the living room coffee table and the writing desk.

Perhaps you enjoy living among junk-filled opulence, but rest assured that it is being made possible by an industrial base that is consuming nonrenewable natural resources and destroying the biosphere at a rapid pace, is financed by accumulating a mountain of unrepayable debt and is enabled by near-slave labor of economic migrants both here and abroad. But if you ever decide to downscale to a point where everything you need to live well can be provided by you yourself, with the help of a few friends, in order to render yourself—what's that word again? Oh yes—*naturelike*... then here is a time-tested plan that works very well.

The stove

The design of the Russian stove is several centuries old and seems to have emerged soon after the spread of firebrick, which is a formulation high in silica that is less susceptible to spalling when heated repeatedly. It is a massive masonry structure with its own foundation. At its center is a vault with an arched ceiling and a flat floor, often high enough for someone to squat inside. Fire is set inside the vault, far inside the stove. At the front of the stove is a flue, which includes a dogleg with a gate that is used for hanging meat and fish for smoking. Right back of the flue is a threshold that protrudes down from the top of the vault, holding hot combustion gases inside the innermost part of the vault, resulting in better heat transfer. The top of the vault is filled with solid fill and covered over with a layer of brick, forming a platform, and a straw-filled mattress, which is often big enough to serve as a bed for an entire family of five. Between October and May, when the stove is fired twice a day, the temperature of the platform stays at a constant, comfortable 25–27°C (76–80°F). During the hot part of the summer, when the stove is not fired because cooking is done at an outdoor hearth, the stove provides a cool place to sleep.

The outer wall of the stove has several niches. They improve heat conduction from the stove to the air in the room and are also used to dry clothes, herbs, mushrooms and berries, to keep food warm and to provide a place for the samovar, which boils water for tea. The firebox of the samovar, typically stoked using pine cones, exhausts into the flue of the stove. Under the stove is a space that is used to store firewood and can be a warm place for animals to sleep. The stove can also be used as a sauna—by sitting cross-legged inside the vault when it is relatively cool.

The Russian stove includes an entire dedicated set of utensils that are specific to it, each perfected over the centuries to have the largest possible set of functions. Food is cooked in clay pots and in cast iron skillets that lack a handle. The pots are placed inside the stove using stove forks, which come in three sizes and grab pots by the neck, while the bread and the skillets are moved about using a flat-bladed wooden spade, similar to the paddles used to handle pizza.

For the sake of comparison, let's consider what you'd have to shop for if you didn't happen to have a Russian stove. To heat the house, you'd need to buy a furnace and either install an oil tank or hook the house up to a gas main. Then you'd need to construct a way to distribute the heat, through either forced air or baseboard heating, and this involves installing lots of either ducts or pipes. You could also install a modern, energy-efficient wood stove, but then the bedrooms would be cold, so you'd probably run out and buy some electric space heaters and, to keep the beds warm, some electric blankets. To cook food, you'd need to buy a cooking stove with an oven, either gas or electric, a toaster and a microwave oven. You'd need a separate smoker for smoking fish and meat, plus some drying racks for drying things. Or you could just get rid of all this expensive, short-lived junk and render yourself *naturelike* by building yourself a Russian stove and using it in place of all of the above.

The sauna

I am not sure how you'd go about shopping for a sauna, which in North America is only included in the wealthiest homes and at gyms and spas, whereas every Russian peasant has had access to one since time immemorial. Even the poorest villages have some number of little log cabins, consisting of two small rooms—a vestibule and an inner room with benches, bunks and a stove that heats a cauldron of water and a pile of rocks. The word "sauna" is a Finnish borrowing; the Russian word for it is "баня" (*bánya*), but the two traditions are very similar.

After water has been brought in buckets and poured into the cauldron, and the furnace stoked and allowed to burn out, people enter the vestibule in twos and threes, undress, enter the inner room, mix hot and cold water in washbasins and soak and scrub themselves clean. Then they sit down on the benches and throw water at the hot rocks to make steam. After steaming themselves for some time, they whip each other with switches made of dried birch boughs, which have just the perfect rough, serrated edges for pealing off the outermost, dead layer of skin. It is also a wonderful way of getting rid of pent-up aggressions: few people can hold a grudge against someone they just whipped with birch boughs in a dark, steam-filled room.

After the sauna, many Russians like to go for a roll in a snow-bank or a dip in an ice hole. The combination of extreme heat and extreme cold, with a sudden transition from one to the other, seems to render the human body quite inhospitable to most pathogens and is in many cases a one-stop cure for the common cold. The sauna is also a potent remedy for lots of aches and pains, from simple muscle pain from overexertion to arthritis, hypertension, poor circulation, respiratory problems and numerous other complaints.

The Russian stove, the Russian log cabin of which it is the centerpiece and the Russian sauna are examples of *naturelike* technology par excellence. They can be constructed in ways that cause zero harm, and their benefits are numerous. Although the stoves are often constructed by skilled craftsmen, this is an artisanal rather than an industrial activity, with numerous local adaptations, while the cabin itself is perfectly vernacular in all of its many manifestations.

Time for a change of venue?

SUPPOSE YOUR SITUATION is such that you need to effect a swift change of venue. The circumstances that prompt this relocation can be quite varied, but here are some of the common and foresee-able ones, based on developments which we can already observe in many parts of the world. Clearly, the idea of fleeing is not one to be entertained without good reason, but if the area where you live is likely to stop being survivable, whereas you would very much like to survive, then that is a very good reason indeed. For example:

1. There is no more fresh water where you live. The reservoirs are dry and dusty, the artesian wells are either no longer producing or are producing water laced with arsenic and heavy metals, while the few desalination plants bottle their water and sell it at prices you cannot afford. What was once field and pasture is eroding to sand dunes. Forests have dried out, burned down, and are now a lunar landscape crisscrossed by deep ravines made ever deeper by sporadic torrential downpours—too sporadic and too torrential to be of benefit to vegetation.

2. The place where you live is under a few feet of ocean water mixed with raw sewage and mats of blue-green algae—not all the time, but often enough that staying there has become a very bad idea. An onshore wind combined with a high tide and a bit of rain are enough to make contaminated, brackish water spew out of every storm drain. With each passing year more and more basements are flooded, more and more foundations undermined, more and more buildings condemned. Places further inland flood more rarely but are already too crowded and will be subject to the same conditions after a slight delay.

3. The place where you live happens to be in the crosshairs of a new kind of superstorm. Glacial melt, accelerated by global warming, has caused a lens of lighter, fresh water to form on top of part of the ocean, stalling ocean currents and causing very large temperature differences between neighboring patches of ocean. This combination of factors has supercharged tropical storms, driving them to a level of intensity not seen since over 100,000 years ago, when they generated waves large enough to pluck giant boulders off the ocean bottom and hurl them atop mountain ridges on shore.[7]

4. Your country has been overrun by migrants who have looted the shops, occupied many of the public buildings and are busy beating up the men and raping the women.[8] There are large sections of your

---

7   http://www.atmos-chem-phys.net/16/3761/2016/

8   Like they are doing in Sweden, which now has the second highest rate of rape in the world, after Lesotho, which is a state within South Africa.

city where even the military, never mind the police, fear to venture. But even the rest of the city is not the least bit safe. Men without proper Islamic beards and women whose heads are uncovered are attacked without warning. Property crimes and home invasions by migrants are not prosecuted for fear of giving them an excuse to start a riot.

5. Your country has gone full-retard fascist. Your best option is to work a soul-destroying corporate job while slowly sinking deeper and deeper into debt, hoping against hope that you will make it all the way to retirement, even as you watch your colleagues being replaced by machines, illegal immigrants and underpaid foreign contractors. Your second-best choice is to subsist on meager social benefits, most of which go to pay for drugs, which you need in order to hold on to what remains of your sanity, while the pressure of perverse government incentives destroys your family and your children gradually turn feral. Whichever option you choose, you are electronically monitored 24/7 and are absorbed into the for-profit prison system for the tiniest transgression, where your best chance to survive is by working as a slave.

6. You are doing fine economically, but you find your environment, both physical and human, increasingly unsatisfactory. Everything you see around you is cheaply slapped together out of industrially produced components, dressed up with a gaudy plastic veneer to make it "look nice." If it all looks computer-generated, that's because it is. All the people around you do their best to ignore the real world, focusing instead on distractions such as television sitcoms and video games. This, for them, is a reasonable choice, since their physical environment is just an older, no longer fashionable version of what they see on the screens of the mobile computing devices to which they are hopelessly addicted. They are obese, emotionally stunted, physically helpless and, as far as you are concerned, it would be helpful if they didn't even exist. In fact, you'd enjoy seeing them replaced with cages of parakeets, potted plants or nice round rocks in a Zen garden. Their parents and grandparents once got things done by pushing buttons on machines, but

now it is the machines that push their buttons and program them to say and feel various things on command. You can't help obsessing over the thought that this is not real life, that real life must be somewhere else and that you must go and find it before you run out of time.

7. Any combination of the above, including all of the above.

Extreme homesteading

I AM QUITE hopeful that there will be numerous pockets around the planet that will remain survivable, but there is one gigantic patch of land that is far enough north and high enough above sea level to avoid being ravaged by rapidly rising global temperatures and oceans, and has a sufficiently healthy and powerful biosphere: Siberia. You are probably not immediately enamored of the idea of moving there, but this is just a case study of what is possible—a thought experiment whose goal is to show what sorts of skills, adaptations and changes of habit would be required in order to survive and thrive in what is probably one of the most challenging environments on Earth. The problems would be much the same elsewhere although, obviously, the further south you decide to settle, the less of a problem it will be to get things done during the summer and to keep warm during the winter, and the more of a problem it will be to avoid death from heatstroke, dehydration, and starvation when crops fail.

For the sake of this thought experiment, let us assume that the logistics and the political situation around your relocation have been sorted out: your papers are in order and you have a berth on a ship that will take you to a river port near your destination. From there, a riverboat will take you upstream to a spot near your assigned 100 hectares (250 acres) of land, where you, together with a small group of like-minded others, will be left alone with enough supplies to make a fresh start. You slip away in the night with just a change of clothing and a pocketful of mementos, quiet as a cat, never to be heard from again.

Your land is being granted to you by the government in the form of a permanent, heritable lease, with no commercial rights over it whatsoever, for you and your children to use sustainably in perpetuity, for as long as you physically reside on the land. The terms are not particularly onerous: you are taxed only on home-produced goods that you sell, and one of your sons may be conscripted in case of a national emergency, provided he is not your only or your eldest son and not a younger son either if he is the family's main provider.

But there is a problem: your land is quite far north. Nine months out of the year, the temperature there is near or below freezing, and during the coldest four to five months it can get as cold as -40°C (equal to -40°F). In the dead of winter there are only three hours of sunlight. But during the three summer months temperatures soar to +35°C (95°F) and there are 21 hours of sunlight. Another problem is that the land is not easily accessible. There are no roads; nor are there plans to build any. During the summer it is accessible on foot and over water; during the winter it is accessible by ski and sled, over snow-covered land and frozen water. During spring, when trails turn to mud and broken ice rushes down streams and rivers, it is not accessible at all. Nor is it accessible during autumn, when snow falls on ground that isn't frozen yet and forms a heavy, wet slush, and when the ice on waterways is already too thick for navigation by boat but not yet thick and solid enough to travel over.

But there is also good news: each year, the climate is getting warmer, with the frosts arriving later, the thaws setting in earlier, the growing season getting longer and more and more deciduous trees taking root in sunny, sheltered spots.

A riverboat will drop you off at the water's edge within less than a day's hike of your land. It will be in early summer, after the rivers are clear of ice and the riverbanks are no longer flooded. You will have just enough time to prepare for next winter so that you can survive it.

What you can take with you is what you and members of your party can carry on their shoulders, ferrying supplies from the river's edge to your plot of land. This basic kit includes:

1. An axe and spare axe heads
2. A knife and several knife blades without handles
3. Shovel heads
4. Saw blades
5. A file for keeping all of these sharp
6. A shotgun and a dozen shells
7. Heavy boots, a parka and other cold weather gear
8. Several changes of clothing per person
9. An emergency medical kit
10. A few pots, cups, spoons, forks
11. A samovar
12. Several sacks of grain (rye)
13. Several sacks of potatoes
14. Assorted seed packets
15. Canvas tents
16. A small assortment of tools (such as a sewing kit) and supplies (such as tea)

You will also be bringing with you a few animals:

1. Dogs (one of them male) to serve as your security system and to help you hunt and pull sleds
2. Cats (one of them male) to keep the rodent population under control
3. Chickens (one rooster) to provide eggs and meat and to keep the bugs under control

This, plus your body, is all of your initial "hardware" which you will use to bootstrap the entire operation; everything else is "software"—and it has to be downloaded directly to your brain before you begin, with a full back-up in somebody else's brain in case something goes wrong with yours. This is your Naturelike Technology Suite (NTS) and if you use it correctly, your chances of surviving, living a long and happy life and leaving behind happy, healthy, self-reliant children are much better than in any and all of the typical scenarios outlined above.

The land is neither farmland nor pasture but boreal forest, thick with coniferous trees, mostly pines and firs. There are plenty of animals you will be sharing it with, especially in the summer when the migratory birds make their appearance and many other animals are about. Your first concern is with bears, which have come out of hibernation some time ago, but are still hungry and very ornery. The wolves may also take a keen interest in your camp. You will need to impress upon all of them that this is now your territory as well as theirs, by keeping fires lit at night, never going anywhere unarmed, screaming at them and physically threatening them whenever you see them and other such measures. You must take up and defend your position at the top of the local food chain. Shooting one alpha male of both the wolf tribe and the bear tribe, even if it means using up a few shotgun shells from your precious collection, skinning them, tanning the furs and sewing them into hats and coats sends an unmistakable message: there is a new apex predator in these woods; act accordingly. As for the rest of the animal kingdom, you should try to make peace with them or let your animals handle them. If you leave them alone and sometimes (but only sometimes, on specific occasions) offer them food, they will become semi-tame over time and will be much easier to catch by setting traps. Of these, back-breaking deadfalls are the most humane.

But your first and primary task is to fell trees—as quickly as possible, propping up the logs in sunny places so that they have a chance to dry out. The time to harvest timber is before thaws set in and the sap starts running, because after that the logs become much heavier and more difficult to work with and move, will not burn as well and will rot much faster if you build with them.[9] But

---

9　This is the exact opposite of what you would do in the tropics. There, you would harvest wood when it is full of sap to protect it against insects and fungi.

you have arrived too late to do that and have to make do with wet, heavy logs.

Your second task is to get food and avoid depleting your supplies, which are for planting, not for eating. A spring thaw is an excellent time to get moose and reindeer, which can't run away because of the heavy, wet snow. Until the ice breaks, ice fishing also remains a possibility, and you can preserve your supply over the warm months by hot-smoking and drying the meat and the fish. But, again, you arrived too late, and your best chance to catch enough food is by setting traps and building weirs to catch fish.

Your third task begins once the ground has thawed out and dried out enough to dig. You need to move out of tents and into a slightly more permanent dwelling before winter. Constructing a log cabin during the first season is out of the question, because there is simply too much else to do and because you arrived too late to get logs that are free of sap. But you can certainly harvest enough logs to build a dugout bunker that will last a few seasons. This is done by choosing a patch of land with good drainage and digging a trench. At the back of it is a hearth, along the sides are bunk beds. The roof is created using a layer of logs, the cracks between them packed with moss and insulated by covering it with a thick a layer of dirt and sod. The hearth should have a flue and a chimney high enough to stick out above the snow, or your fire will keep getting extinguished by meltwater. Two doors with a vestibule between them are an excellent idea. The vestibule will be used to store your supplies of frozen meat. The doors must open in rather than out, or you will be trapped inside by snowdrifts.

Your bunker should be surrounded by a wicker fence, constructed by driving stakes into the ground at intervals and filling the spaces between them with tightly packed twigs or saplings. Fence in an area that is round or oval rather than square, for a 25 percent increase in the amount of area enclosed by the same length of fence. A round fence also makes it easier for your animals to catch interlopers because there are no corners where they can hide and burrow. Curved fences are also better at resisting wind and snowdrifts.

Your fourth task will be to grow food. The land you've cleared by chopping down trees is covered with a thin layer of poor forest soil, acidic because of all the pine and fir needles, and is not immediately useful for planting. But if you dig various things into it, you will be able to use it to grow all of your staples: potatoes, rye, cabbages and turnips. Ashes from the hearth, thoroughly rotted tree trunks and mud dredged out of nearby streams all make useful soil amendments. Potatoes can be planted as chunks containing eyes, or buds, with one or two eyes per chunk, and the rest of the potato can be eaten. Rye can be grown in quite poor soils and is amazingly stubborn and keeps growing until it goes to seed. Because of the nearly 24-hour sunlight and warm temperatures, everything will grow very fast. Your animals will be kept busy and well fed by all the moles, voles, mice, slugs and snails that will be trying to eat your produce before you do.

By the time you are done growing and harvesting the food, days will start getting shorter and by sunrise frost will appear on trees and the walls of your tent. It will be time to move inside your bunker and start heating. Before the migratory fowl fly away, be sure to get some geese or, failing that, ducks, and save their fat for the winter. To avoid frostbite, smear goose fat on any exposed skin when you venture outside in the dead of winter.

Once the temperatures are reliably below freezing, but before the winter blizzards set in, try to stockpile as many animal carcasses as you can, to gradually hack away at and defrost as the winter wears on. This is the time of year when animals are at their fattest and most complacent, and those that are the oldest and the least likely to survive the winter are ripe for the picking; if you don't get them the wolves will. Dietary fat, as a source of calories, is particularly important: in a cold climate, it is almost impossible to get enough calories to stay warm while working outside in any other way, and how much winter work you will get done will be directly determined by how much animal fat you can get your hands on. Don't worry, eating fat will not make you fat; the fastest way to get fat is by eating processed carbohydrates and refined sugar, and you won't have that problem.

At the beginning of winter, most of your work outside will involve cutting, splitting and stacking firewood from the logs you harvested in the springtime, since you do not want to be out swinging an axe when it's -40°C outside and blowing a blizzard. Once your supply of firewood is laid in, there are other tasks to attend to.

First, you need to get serious about trapping for fur. The parka you brought with you will wear out and will need to be replaced with a fur parka you will need to sew yourself. The animals you trap will be frozen solid by the time you get to them and can stay that way until springtime. You can gut them and skin them when they thaw out, saving the brain and the liver for tanning the pelt. The pelts will also serve as valuable trade goods—about the only ones you will be able to come up with during the first few seasons—and you will need trade goods in order to barter for the supplies you will need.

Second, if you are close enough to a river or a lake to make it there and back by daylight, you might also attempt some ice fishing, although without skis and a sled (unless you found time to make them already) your range will be quite limited.

Other than that, most of what you will do during the winter is cook, feed yourself, feed the animals, drink tea, tend the all-important fire and sleep a lot. The tea is important because working outside in cold temperatures is extremely dehydrating: the cold air sucks the moisture right out of you. This is why a samovar (which is stoked using pine cones or wood chips) is included in your initial survival kit. Trying to boil enough water in a pot over a hearth is far too slow and rather inefficient. But a bucket hung over the hearth is quite useful for melting snow to get water for drinking and washing without having to go anywhere to fetch it.

Before spring thaw arrives, you will need to get busy harvesting logs—for next winter's firewood as well as for building the log cabin. Once that's done, you will have won, surviving the most difficult first season without starving or dying of exposure and ready to build your homestead. Once that's done, you will be well on your way to making a perfectly reasonable life for yourself and your family using the rest of your NTS.

• • •

MOVING ALONG WITH our case study/thought experiment in extreme homesteading, let us assume that you have survived your first winter on the land. Congratulations! The worst part of the ordeal is quite possibly over. Gone are whatever addictions you had on arrival, be they internet access or coffee. Your new world consists of the few people around you and a much larger number of plants and animals. But it is a world that is indisputably yours—to make the best of and to pass along to your children and grandchildren.

In the beginning some elements of unnaturelike technology will persist. But as the seasons wear on, your newfound world will come to no longer include electricity or electronics, synthetic materials or fabrics, internal combustion engines (no more outboard engines, snowmobiles or chainsaws), firearms, synthetic pharmaceuticals, biotechnology or much else. All of these will quietly fade from memory. Their loss will be quite inconvenient for you, to be sure, at least initially, but good for nature. Many of the things you previously accomplished using power you now accomplish just as quickly using a combination of clever timing, stamina and technique. For example, having no internal combustion engines or draft animals to transport logs, you cleverly arrange for them to be carried downstream by the spring flood and conveniently form a logjam close to your homestead.

In place of gadgets there are printed books: the riverboat that makes its rounds of shoreline settlements at most once a year—in midsummer—carries a lending library, dropping off books one summer and picking them up the next. It also distributes a set of textbooks made available by the government: language and literature, mathematics, botany, biology, chemistry, physics, geography and geology. Some of the textbooks haven't changed in many generations; after all, there have been very few recent scientific discoveries that would be useful to you. Others have needed an update or two; for example, the geography textbook no longer lists countries such as Bangladesh, Kiribati or US states such as Louisiana and Florida, which haven't been heard from in quite a while. Numerous failed states with morbid populations and unguarded borders will be given scant mention.

In place of synthetic fabrics or cotton there is cloth of flax and hemp (cotton went away along with industrial chemistry on which it depended for pesticides). You make much use of leather, wool and fur, the last of which is already essential for your continued survival. When in the spring your dogs start to shed, you brush them out and use the dog hair, which has amazing curative properties, to knit socks, mittens and scarves. Since pharmaceuticals are largely gone, everyone is busy gathering and cultivating medicinal plants and practicing preventive therapies. A favorite for killing off viruses is a trip to the sauna followed by a roll in a snowbank or a dip in an ice hole.

Metals are about the only relic of industrialism still in widespread use. There is no practical limit to the amount of mild steel scrap that is available from industrial ruins—enough to keep all the blacksmiths (of a much smaller and widely dispersed population) busy for thousands of generations. Copper remains a favorite, since it can be cold-worked into any shape.

. . .

THIS MAY SEEM like a harsh life, but all of the alternatives will be even worse. As the average global temperature rises by over 15°C— far in excess of the 2°C still bandied about by the politicians and their court scientists at the Intergovernmental Panel on Climate Change (IPCC)—most of the inland areas further south will be made unlivable by summer heat waves with wet bulb temperatures in excess of 35°C. Without air conditioning such temperatures are lethal, and summer heat waves, accompanied by blackouts, will kill off entire regions. Coastal cities will perish for a different reason: ocean level will rise by as much as 120 meters, putting them permanently under the waves.[10] With the disappearance of mountain glaciers entire countries that depend on glacial melt for irrigation—and there are many of them—will starve. For populations used to living on the coasts and earning a living from the sea moving further inland will not help much—because of all the nuclear power plants that will go underwater with their spent fuel pools still stocked, producing hundreds of new Fukushimas that will make the oceans too radioactive to fish. And as climate change continues and accelerates, all of these problems will grow progressively worse.

But then here you will be, near one of the major Eurasian north-flowing rivers that empty into the Arctic Ocean—Lena, Ob' and Yenisey.[11] You are high enough above the quickly rising ocean level and away from everything else—including the few still crowded population centers that will be getting ready to go through an episode of mass die-off. If the summers get too hot or too dry, you will be able to build rafts and transport yourself further downstream, closer to the Arctic Circle, where it will be cooler and wetter. All the while, you can go on practicing your Naturelike Technology Suite, some of which has not changed much since the landscape you now occupy was first settled thousands of years ago. In the summer, the now ice-free, navigable Arctic Ocean will allow the surviving remnants of humanity to keep in touch.

---

10   See David Wasdell's report at http://apollo-gaia.org/harsh-realities-of-now.html. Wasdell concludes that "With an expected total cumulative carbon emission of around 2000 Gt C [we should expect a rapid global average temperature increase of] around 10°C, with the extension to full equilibrium response of more like 15°C. An ice-free world and a sea-level rise of around 120 meters are in prospect."

11   Previously, the McKenzie River in Canada could have made a promising destination, but recently its headwaters have become badly contaminated by the mining of tar sands.

Life on the move

IN THE PREVIOUS case study/thought experiment, we took a close look at what it would take to survive the future we've guaranteed ourselves on any one patch of land. But I would be remiss if I didn't mention an alternative: while there will be a dearth of places that will be survivable year-round, it is likely that many more opportunities will exist for lifestyles that are either nomadic (wandering from place to place) or migratory (with semi-permanent seasonal camps). These lifestyles come with their own naturelike technology suites—which are much more challenging than the ones required for a settled lifestyle, simply because a mobile, portable technology is more demanding than a fixed installation.

Doing away with the a fixed abode confers numerous advantages: you become free to move to flee danger; you are prevented by your circumstances from wasting your energies on accumulating possessions beyond those you absolutely need and use all the time; you get a chance to construct your own shelter to suit the situation. These are all practical considerations, but there is more to being nomadic than just being practical. Nomadism, you see, is not just a good adaptation for uncertain times. It is also godly and sublime.

Most people, when they hear the biblical phrase "the house of the Lord," imagine a cathedral or a temple. Their fixed notion of a house is a large, permanent, immobile structure. What a surprise it is, then, to learn that the house of the Lord was, to begin with, most definitely a tent: Ancient Hebrew *beth* or Arabic *beyt* are both words that signify "tent." The tension between the settled and the nomadic is present throughout the Bible. It is the tension between slavery and freedom, and the biblical account makes it clear that God or Yahweh—who was originally a nomad god, the Bedouin god of flocks and herds—always sides with the nomads.

Let's look back at one of the world's great founding myths, the story of Abraham, who gave his name to the Abrahamic religions of Islam, Judaism and Christianity, whose adherents account for more than half of the population of the Earth. In the story, Abraham and Lot, his nephew, leave the city and with their herds travel to Canaan and live there as nomads at the edge of the desert. But they quarrel, and Lot departs for the cities of Sodom and Gomorrah. Yahweh punishes him for his choice, destroying the cities and turning his wife into a pillar of salt just for looking back at the destruction, while Abraham stays pure and on the move, and his two sons, Ishmael and Isaac, live on to create the two great nomad tribes, the Arabs and the Jews.

Although nomadism is the ideal, the tension between the nomadic and the settled is ever-present. Droughts, famines and political oppression often force nomads to take refuge among the settled. If they stay long enough, they may lose their nomad ways and become stranded. Even Abraham was driven by famine to leave Canaan and take refuge in Egypt for a time, but was quick to escape as soon as conditions improved. Later, another famine forced his descendants back into Egypt and a life of servitude, but here their sojourn among the settled lasted long enough for them to lose their nomadic skills, condemning them to a life of slavery. But they managed to produce a visionary—Moses—who married a Bedouin woman. This woman, Zipporah, daughter of a shepherd, turned out to be the key cultural transplant that allowed the Jews to escape from captivity back into the wilderness and regain their freedom.

Nomadism is culturally and technologically advanced, requiring such elements as portable shelter, a relationship with animals that borders on symbiosis and the ability to self-organize in groups large and small, to survive in a harsh and nearly barren terrain and to control and defend a large and ever-changing territory. In all nomadic cultures, more than half of this cultural and technological DNA is the explicit domain of women, for it is the women who create and maintain the "house of the Lord"—the tent. Men practice animal husbandry, make tools, hunt, fish, fight, spin yarn and make tent poles, but it is the women who weave and stitch. The tent is typically part of the dowry and remains in possession of the woman, hers to keep in case of divorce.

Walk into the tent of any nomad anywhere in the world, from the tropical deserts to the high Arctic, and you will find the same separation of male and female concerns reflected in the interior layout. To the left of the entrance is the women's side. Here, stacked along the walls you will find everything needed for preparing food, for working with leather and fabric and for taking care of children. To the right is the men's side. Here, stacked along the walls you will find tools, weapons, saddles and harnesses. In the middle is the hearth; to the back of the hearth is the sacred place, with an altar. Before the altar is the seat of honor. In case of the Arabs, the separation is enforced using a curtain called the qata, while in the tipi of a Native American the separation is implicit, but it is always there—a nomadic cultural universal. This evolved trait makes perfect sense: the life of the nomad is so complex and requires such a high level of competence that a separation of concerns between men and women is essential to survival. A lone male can perhaps lead a nomadic existence until he dies, but for nomadism to exist as a viable culture requires a woman-nomad, with woman-nomad know-how.

Women tend to be more conservative than men (politics aside) in that they tend to pass on their skills to their daughters more or less unchanged. Thus we find, in nomadic architecture, a striking stability of forms. The black tents described in the Bible, under which the Israelites camped in the Canaan desert, are still to be found along a desert belt stretching from Casablanca on the Atlantic coast of Africa all the way to Tibet (where they use the long belly hair of the yak to weave the fabric). It is a rectangular piece of goat-hair fabric, stitched together out of wide woven strips, erected using a few poles and stretched using long lines secured to pegs. It keeps the interior cool by blocking sunlight and by creating an updraft that pulls air up through its loose weave, but when it rains the goat hair fibers swell up and create a waterproof surface that sheds water.

North of the black tent belt runs the yurt belt. Yurts use a free-standing frame that consists of a barrel-shaped latticework at the base, a tension band at the top of the latticework, a crown, sometimes supported by center poles, and poles which are mortised into the crown and hooked onto the tops of the latticework. Over this frame is pulled a covering of felt, its thickness in proportion to the coldness of the climate. A fair percentage of the population of Mongolia lives in yurts to this day, and yurt-dwelling Mongols once made it as far west as the gates of Vienna. Buckminster Fuller's dymaxion house was essentially a yurt—fabricated out of aluminum, which is an unfortunate choice of material since it doesn't grow on trees or on sheep.

North of the yurt zone and throughout the circumpolar region we find two basic shapes: the cone tent and the dome tent, covered either with skins and hides or with steamed birch bark. Inside, we often find the same layout: hearth in the middle, women to the left, men to the right, altar in the back. The Koryak-Chukchi *yaranga* is particularly notable. These tribes, which inhabit the very farthest north of Siberia, use a tent within a tent, called a *polog*, to keep warm in spite of temperatures that are often colder than -40°C below. The inevitable condensation is dealt with by taking the *polog* out during the day, allowing the condensation to freeze solid and beating it out with a stick.

Nomadism is an innovation, requiring a great deal of advanced technology and know-how. It is relatively recent, and in many places its advent coincided with the domestication of various animals. It is the symbiosis with these animals that gave the nomads their speed, range and ability to sustain themselves in places where a stationary population would quickly perish of hunger and thirst. In the desert, black-tent nomads rely on the camel and, in the case of Tibet, the yak; the yurt nomads of the plains rely on sheep and horses; the circumpolar tribes rely on the reindeer in Eurasia and its undomesticated cousin the caribou in North America. Prior to the advent of nomadism most of the places where only nomads could survive remained uninhabited.

There will be places in the world where not even a nomadic tribe will be able to survive, but they have the option of moving on when they see circumstances changing for the worse, while settled tribes lack the know-how to do so. Settled populations depend on a stable climate to be able to bring in crops from the same patch of land season after season. Over the past 11,000 years this was possible in many more places on Earth because during this period of time the climate was particularly stable and benign, but it appears that this period is now over. The Earth has entered a period of climate upheaval in which the regular patterns of nature on which agriculture relies can no longer be taken for granted. Under these conditions, settled populations in many parts of the world will quickly run out of options.

Although the cultural preference in many parts of the world has been to disrespect the nomad, it is likely to turn out, for more and more people, that their choice lies between turning nomadic (if they can) and perishing in place. And it bears repeating that being nomadic requires a much higher-level set of technologies than just staying in one place—one that may be difficult to perfect in a single generation or even in a single lifetime.

6
POLITICAL TECHNOLOGIES

YES, IF WE put our minds to it, we just might make it: find a relatively intact bit of biosphere and come up with a set of naturelike technologies that will allow us to survive within it, for many generations, without causing it harm.

But there is another problem, and it's a very serious one: people won't let us. Wherever we go, some entity that is a projection of the technosphere—either a very wealthy individual or a corporation—claims to own the ground under our feet and attempts to exercise control over how it is used. And what that entity typically wants to do is to make it turn a profit even while destroying it. Even if you manage to find a patch of wilderness that is not under anyone's specific control, you will no doubt find that ensconcing yourself in it with the goal of living in balance with nature by using naturelike technologies is abhorrent to civilized society, which will send people after you to drag you back to civilization.

This is a political problem. In an ideal political environment, the masses of people would realize what needs to be done, hold a referendum and that would settle the issue. Wealthy interests and corporations would be dethroned, the technosphere would shrink, and a new way of inhabiting the landscape based on naturelike technologies would quickly take hold. But instead we live in a

world where most people's minds are spellbound by a set of political technologies, and their political self-expression is funneled
through electoral systems which are rigged in such a way that even
if the spell were suddenly broken and everyone voted for a survivable future for their children, you can be sure that their votes
would be thrown away.

But there is some hope, for some people and certain places, that
this can be changed. The understanding of how these political technologies function does exist, and it can be put to use in breaking
the spell and in undermining the political machines. There is also
an inkling of an idea for how to arrive at a political arrangement
that would be beneficial and conducive to the goal of shrinking the
technosphere.

Beyond good and evil

POLITICAL TECHNOLOGIES HAVE three main goals:
1. Changing the rules of the game between participants in the political process
2. Introducing into the mass consciousness new concepts, values,
   opinions and convictions
3. Manipulating human behavior directly through mass media and
   administrative methods

Political technologies pursue these tactical goals in accordance with higher, strategic imperatives, and it is only the noble
nature of these higher imperatives that can justify the use of
such high-handed, antidemocratic means. Yes, the ends do
justify the means—once in a great while: it is unequivocally
better to save humanity and the natural world through nondem-
ocratic means than to let it be destroyed while adhering to strictly
democratic ones.

But often the imperatives are far less than noble. They can be
separated into two kinds:

I. To improve everyone's welfare by pursuing the common good of the entire society, as it is understood by its best-educated, most intelligent, most levelheaded, decent and responsible members. Political technologies of this kind result in a virtuous cycle, building on previous successes to increase social cohesion and solidarity, and setting the stage for great achievements. (These are the good kind.)

2. To enrich, empower and protect special interests and privileged elites at the expense of the rest of society. These kinds of political technologies either fail through internal contradiction or result in a vicious cycle in which those who benefit from them strive for ever-higher levels of selfish behavior at the expense of the rest, setting the stage for injustice, exploitation, poor social outcomes, economic stagnation, mass violence, civil war and eventual political disintegration. (These are the bad kind.)

Political technologies in the US

LET'S CONSIDER THE United States, which offers a particularly spectacular example of political technologies at work. The United States currently has more than its fair share of them, especially of the bad kind. Let's briefly review a dozen of the most important ones.

The fossil fuel lobby

OBJECTIVE To convince the US population that that there are no viable alternatives to burning ever-increasing quantities of fossil fuels and that catastrophic anthropogenic climate change is not occurring.

MEANS Smear campaigns against climate scientists, injection of fake science, denigration of science as a whole, portrayal of the movement to stop catastrophic climate change as a conspiracy, etc.

This example alone is sufficient to illustrate how effective a political technology can be: we all know plenty of its victims.

Even quite intelligent people often espouse the opinion that the observed climate change is the product of natural variability (it isn't) or that the efforts to mitigate climate change are a conspiracy of a world government (which doesn't exist). This clearly shows how effective political technologies are: they can warp the minds of both the simple and the intelligent. Although they can also be used to unwarp that which has been warped, there are, unfortunately, no examples of political technologies within the US that have been used in pursuit of the common good. With regard to what passes for politics in the US it is best to avert your gaze until its glowing embers cool because you'd be burning your eyeballs for nothing.

American political technology shows some signs of failing through internal contradiction. For example, in the late summer of 2015 parts of South Carolina—a self-styled "conservative" state—went underwater in a so-called "1,000-year flood" (soon to be renamed a "100-year flood," then a "ten-year flood" and, finally, "blub-blub-blub"). Unlike North Carolina, Florida (another "blub-blub-blub" state) and Wisconsin, the South Carolina legislature hadn't banned the use of the term "climate change" by state workers; not that anyone has, in any case, ever heard them use this term. When political technologists start banning the use of certain words, you know that they are becoming desperate. When a political technology shows signs of failing through internal contradiction, it is often best to let things run their course. After all, what does it matter whether officials in the Carolinas or in Florida use the term "climate change" or the term "blub-blub-blub" as they disappear under the sewage-contaminated waves?

The arms manufacturers

OBJECTIVE To convince the US population that private gun ownership makes people safer, is effective in preventing government tyranny and is a right to be defended at all costs. This too is showing some signs of failure through internal contradiction, as the number of mass shootings in the US, as it were, shoots up. But the level of brainwash here is rather high, and the US authorities may find themselves forced to resort to direct manipulation to bring the situation under control—or, as is becoming increasingly likely, fail to do so. This may involve some sort of mass standoff between the government and the "gun nuts," in which the gun nuts are declared to be terrorists, outlawed and, in a demonstration exercise, simultantaneously wiped out by the army, the navy and the air force. But this would only bring out the next layer of internal contradiction; by decisively demonstrating that owning a gun does not make you safer and that guns are useless in preventing tyranny, the government would be forced to tacitly admit that it is, in fact, a tyranny, at war with its own people. And this would undermine a number of other political technologies on which the government depends for its political survival, such as the one that keeps people convinced that the US government is a force for good in the world rather than a hornet's nest of self-serving special interests.

The two-party political system

This includes not just the two political parties but also the lobbyists and their corporate, big-money and foreign sponsors.

OBJECTIVE To keep the people believing that the US is a democracy and that people have a choice. On the one hand, this technology seems to be working. A lot of the people voted for Obama (some of them even twice!) and then had a difficult time facing up to the fact that he is an impostor, barely different from his predecessor, and that everything he had said to get their vote was just hopeful but duplicitous noise. And just as I am writing this a lot of these same people are ready to vote again—for some other democratic career politician making similar kinds of duplicitous noise. On the other hand, this piece of political technology seems to be in rather sad shape. The party machinery seems unable to produce viable candidates. Moreover, most of the voters no longer identify with either party. This must be an unnerving development for political technologists in charge of herding them toward one end of the political spectrum or the other by dangling unimportant but divisive social issues before them while blocking all dissent on any of the important questions of public policy. A case can be made that this political technology is currently entering a death spiral. Increasing portions of the electorate are supporting candidates from outside the two-party duopoly. In attempting to counter this threat to its hegemony, the political establishment becomes unable to hide the fact that it is a unified front directly opposed to the popular will. This, in turns, makes it even less likely for the electorate to continue supporting the establishment.

Defense contractors and the national defense establishment

OBJECTIVE To justify exorbitant defense budgets by claiming that they keep the nation safe—because they make evildoers everywhere beware or some such nonsense. The US has a very expensive defense establishment but a highly ineffectual one. Case in point: in 2015, when Russia came to the defense of the embattled Syrian government and the hostilities in Syria threatened to escalate, the US ordered the aircraft carrier USS Theodore Roosevelt out of the Persian Gulf, leaving the area without a US aircraft carrier for the first time in six years. The reason is simple: although very expensive and impressive-looking, American aircraft carriers are only effective against poorly armed and disorganized adversaries. When it comes to major powers such as Russia, China and Iran they are no more than sitting ducks, being defenseless against attacks by supersonic cruise missiles, supercavitating torpedoes and anti-ship ballistic missiles, which the Americans neither possess nor can defend themselves against. The relative impotence of American high-tech weaponry against adequately equipped adversaries, coupled with the inability or the unwillingness to deploy ground troops (after the great "successes" in the meat grinders of Iraq and Afghanistan), has produced an erstwhile superpower whose ability to project force is rather circumscribed.

Such obvious signs of weakness (and there are many others) undermine the claim that defense dollars are money well spent. After a time, the message is bound to sink in that the US defense establishment is a self-serving public money sponge that produces useless military boondoggles and baseless, dreamed-up intelligence reports, resulting in a serious internal contradiction. Why, then, does it have to cost so much? Defeat can be acquired for a lot less money. The latest US initiative to drop palettes of small arms ammunition and hand grenades into the deserts of northern Syria, hoping that "moderate" terrorists will find them and use them against the Syrian government, is starting look like a sign of desperation.

• • •

THE LIST GOES on but, for the sake of brevity and as an exercise for the reader, I will let the reader fill in the details about the remaining examples of bad political technologies that are found in the US. Information on them is not hard to find. Ask yourself if each of these political technologies is likely to fail, whether through internal contradiction, by triggering a wider conflict or by causing widespread degeneracy in the population it afflicts.

The medical industry
OBJECTIVE To keep people convinced that private health insurance is necessary, that exorbitant medical costs are justified, that socialized medicine is somehow evil, and that they are getting good quality medical care—in spite of all evidence to the contrary

The higher education industry
OBJECTIVE To keep people convinced that higher education in the US is a good value in spite of its exorbitant costs, the student debt crisis and the fact that over half of recent university and college graduates have been unable to find employment that is in any way related to their degree

The prison-industrial complex
OBJECTIVE To keep people convinced that imprisoning a higher percentage of the population than did Stalin, mostly for nonviolent,

victimless crimes, somehow keeps people safe, in spite of there being absolutely no evidence of that

The automotive industry

OBJECTIVE To keep people convinced that the private automobile is the hallmark of personal freedom while denigrating public transportation, in spite of the fact that if you factor in all of the costs and the externalities of private cars (including such things as the retarded development of children due to carbon monoxide poisoning caused by sitting in traffic) and translate them into the number of working hours it takes to pay for all that expense and damage, and then add that to the hours spent driving, driving a car turns out to be slower than walking[1]

The agribusiness industry

OBJECTIVE To keep people convinced that a diet made up of cheap, chemical-laden, industrially produced food is somehow acceptable in spite of the high levels of obesity, heart disease, diabetes and other ailments which are the result. In addition, to try to hide certain facts, such as the fact that Monsanto's Roundup™ (glyphosate) weed-killer is a known carcinogen

The financial industry

OBJECTIVE To keep people convinced that their money is safe even as it disappears into an ever-expanding black hole of unrepayable debt, and that it is safer to keep their savings in banks (where it earns no interest but could instantly disappear during a "bail-in") than under a mattress

Organized religion

OBJECTIVE To keep people convinced of the importance of kowtowing to a big white man in the sky who might send you to hell in spite of the fact that he loves you; who, in spite of being

---

1   This point is persuasively argued by Ivan Ilich in *Energy and Equity*.

all-powerful, always needs your money; and whose written word takes precedence over using your own reason and relying on facts to find your own way in the world. To cause simple-minded people to insist that a worked-over story of the Egyptian god Horus, stuck together with bits of the Gilgamesh Epic and other ancient myths, is the word of God and the absolute literal truth. To keep alive the fiction that religious people are automatically more moral or more ethical than nonreligious people

The legal system

OBJECTIVE To keep people convinced that the legal system produces justice instead of selling positive outcomes to the highest bidder, that feeding a huge army of well-paid lawyers is somehow worth the money, and that obeying a codex of laws so voluminous and so convoluted that it is completely incomprehensible to the average person—as well as to most lawyers—is what it means to be a good citizen

• • •

AS YOU SEE, the US carries quite a parasite load of bad political technologies. This situation is likely to only get worse because any well-funded special interest group can hire expert political technologists to put together a system that will allow them to get a disproportionately large piece of the pie.

This is bad enough, but bad political technologies cause an additional problem: they debilitate the minds of those they afflict. Their main objective is to keep people convinced of things that are false. Once they succeed, the targeted population becomes personally invested in these falsehoods, comes to identify with them and regards any information that contradicts them either as a personal affront or, at the very least, as a source of unwelcome cognitive dissonance. This makes them impervious to good political technologies—ones that seek to convince them of things that are true and of approaches that do in fact work and steer them in the direction of doing what is necessary. They become what Andy Borowitz called "fact-resistant humans."

Because of its high parasite load of bad political technologies, the population of the US may not be worth the trouble when it comes to putting together good political technologies, such as those designed to prevent catastrophic climate change. A lot of these bad political technologies are destined to fail, either through internal contradiction, by exhausting themselves in futile conflict, or because of their deleterious effects on the people held under their spell. It may make sense to just wait. The logjam of bad political technologies is bound to break at some point, and then we will witness the emergence of an America none of us will recognize.

American political technologies abroad

WE WILL NOW turn our attention to political technologies that have been used by the US against the rest of the world. This may seem like a digression from the task of addressing the question at hand—how to bring about social change in order to avert biosphere destruction, climate catastrophe and human extinction—but it is necessary. In order to shrink the technologies that limit our autonomy, self-sufficiency and freedom, we must first consider those technologies that are specifically designed to *prevent* us from doing just that.

The long list of political technologies used within the US to keep Americans fooled helps us see just how pervasive and destructive they are. What we have not seen are ways to neutralize these technologies—because Americans have so far failed to do so. To find examples of successful ways to neutralize them, we have to look at what the US has been attempting to do to the rest of the world—and failing.

No matter how good America's luck has been—isolated geographic location, plentiful natural resources, the gigantic windfall of its victory in World War II, the additional windfall of the largely self-inflicted Soviet collapse—the luck was bound to run out eventually. In fact, to a large extent it already has: as a purely practical matter, it simply isn't possible to continue running roughshod over the entire planet if you run roughshod over your own population as well. The US has very serious problems besetting its population: half of Americans are obese, a third are on drugs and a quarter are mentally ill. It leads the developed world in deaths from gun violence, police murders and prison population. Half the children are born into poverty and a third into broken and nonexistent families. Over a quarter of the working-age population is permanently out of work. By no stretch of the imagination does this describe a group that can continue to dominate the planet.

Beyond the simple matter of all good (or, if you prefer, evil) things eventually coming to an end, the rest of the world has evolved some effective antibodies against American political technologies, and some of them may be helpful in bringing about the rapid social changes that are needed in order to avert biosphere collapse and climate catastrophe. Before the US empire is swept away in a wave of senseless violence, confusion and embarrassment, we may be able to extract some useful lessons from it.

• • •

WE CAN DIVIDE the political technologies the US uses against the rest of the world into three broad categories. Although the first two may not involve overt, physical violence—at least not every time they are applied—all three categories are actually forms of warfare—hybrid warfare.

International Loan Sharking

John Perkins describes International Loan Sharking in his *Confessions of an Economic Hit Man*:

> Economic hit men (EHMS) are highly paid professionals who cheat countries around the globe out of trillions of dollars. They funnel money from the World Bank, the U.S. Agency for

International Development (USAID), and other foreign "aid" organizations into the coffers of huge corporations and the pockets of
a few wealthy families who control the planet's natural resources.
Their tools include fraudulent financial reports, rigged elections,
payoffs, extortion, sex and murder. They play a game as old as
empire, but one that has taken on new and terrifying dimensions
during this time of globalization.

These efforts eventually produce a bankrupt country that is
unable to service its foreign debt. Whereas in previous eras the US
used gunboat diplomacy to extort payments from deadbeat countries, in a globalized economic environment this has been rendered
largely unnecessary. These days the simple threat to refuse to continue providing liquidity to a country's banks is enough to make
it capitulate. In turn, capitulation leads to the imposition of austerity: public health, education, electricity, water and other public
services are either cut or privatized; public assets are bought up
on the cheap by foreign interests; private savings are confiscated
to make symbolic payments against a ballooning debt to the benefit of foreign banks; subsidies and tariffs are changed to benefit
the rich nations to the poor country's detriment; and so on. Society
crumbles; young people and those talented or educated try to emi-
grate, leaving behind the destitute, the old, the hopeless and the
social predators.

This political technology has worked a great deal of the time,
most recently in Greece, Portugal and Ireland. But there are still
some countries which, although integrated into the global economy, are politically able to withstand this juggernaut and insist
on maintaining their sovereignty and on pursuing a set of policies independent from Washington's dictate. In these cases, the
US deploys a different political technology, which goes under the
name Color Revolution. This technology uses large groups of nonviolent protesters to produce social disorientation, disorganization
and disintegration, to render the political elites within the target
country impotent and to exploit moments of chaos and confusion

to install a puppet regime that can be controlled from Washington, commencing wholesale looting of the country's national assets.

The Color Revolution Syndicate

The methods of Color Revolution are often touted as a nonviolent way to bring about regime change. Gene Sharp, the great theoretician of nonviolent revolution, insists that all protest should be nonviolent. But the concept of nonviolence, comforting though it is to delicate minds, needs to be set aside—because it just plain doesn't reflect reality.

Just because a crowd isn't throwing Molotov cocktails at police while illegally blocking access to public buildings does not make it nonviolent. First, the use of a crowd for a specific purpose is already a use of force. Second, if the demonstration is illegal and if restoring public order requires the use of force, then the crowd is using the threat of violence against itself as a weapon against the rule of law. Calling such a crowd nonviolent is tantamount to declaring that a man making demands while pointing a gun at his own head isn't being violent simply because he hasn't pulled the trigger yet.

The architects of regime change insist on the use of "nonviolent" tactics specifically because they pose a much thornier problem for the authorities than an outright revolt. If the government faces an armed uprising, it knows exactly what to do: put it down. But when the youth of the nation parades around in matching T-shirts (that have been mysteriously shipped in from abroad) shouting deliberately anodyne, aspirational slogans and the entire happening takes on the air of a festival, then the government is unable to act decisively and its ability to maintain public order gradually melts away.

When the conditions are right, the regime changers fly in the mercenaries with the sniper rifles, carry out a public massacre and blame it on the government. These snipers appeared in Egypt in 2011 during the effort to topple Hosni Mubarak. They also appeared in Vilnius in 1991 and in Moscow in 1993. In Tunisia in 2011 they actually got detained. They had Swedish passports and northern

European faces. They said that they were there to hunt wild boar—with sniper rifles, in Tunis.[2]

Let us not allow ourselves to be misled: all three types of political technologies the US uses against the rest of the world are types of warfare—hybrid warfare—and "nonviolent warfare" happens to be an oxymoron. "Nonviolence" is a misnomer; with respect to Color Revolutions, a more accurate term might be "delayed use of violence."

What transpires in the course of a Color Revolution is typically as follows:

GROUNDWORK

The action is instigated by a small, ideologically and politically unified, networked group of elite individuals sponsored by Washington NGOs and think tanks backed by the US State Department. They are often assisted by political and economic accomplices, who generate public unrest through various means, such as imposing economic sanctions or manipulating financial markets.[3] Their goal is to appear to the government as "the voice of the people" and to the people as "the legitimate authorities." They use methods of information warfare: hunger strikes, small demonstrations, speeches by dissidents and symbolic clashes with police in which the protesters play the victim. To hide the fact that they are a small, closed clique of outsiders and foreigners in Washington's pay that has conspired to overthrow the government, they merge into large popular groups of citizens, infiltrate legitimate protest movements and inject their specific slogans alongside popular public demands. Once they achieve a "virtual majority" and accumulate enough followers to march them out for a photo shoot, so that

---

2   An alternative version states that these were real Swedish wild boar hunters who were forced to travel to Tunis by a wild boar shortage in Sweden.

3   A recently failed attempt at Color Revolution in Armenia was instigated by Western accomplices who privatized power plants, then hiked electricity rates, blaming the increase on Russia.

Western-controlled media outlets can champion them as a huge, popular protest movement, they move on to…

THE DESTRUCTION OF PUBLIC ORDER

During this phase, the goal is to achieve maximum social disruption through "nonviolent" means. Streets and public squares are occupied by almost perfectly peaceful crowds of young people chanting moderate, popular slogans. They start by holding officially sanctioned demonstrations, then begin challenging the limitations imposed on them by spontaneously changing the route or by holding meetings longer than scheduled. They start using ploys such a sit-down demonstration accompanied by the announcement of an indefinite hunger strike. While doing this, they actively propagandize the riot police, demanding that the police become "one with the people" and trying to force them to become complicit in what are at first minor transgressions against public order. As this process runs its course, public order gradually disintegrates.

During this phase, it is important that the protesters do not engage in any sort of meaningful political dialogue, because such dialogue may lead to a national consensus on important issues, which the government could then champion, restoring its legitimacy in the eyes of the people while sapping the protest movement of its power. The regime changers pursue the opposite strategy of delegitimizing the government by proliferating all sorts of inconsequential councils and committees that are then held up as democratic, and therefore legitimate, alternatives to the government.

The time of elections is a particularly opportune moment for the regime changers to exploit by claiming that there has been fraud at the polls and by using the social organizations they have infiltrated as fronts in order to claim to be speaking on behalf of the true majority. The White Ribbon Revolution in Bolótnaya ("Swamp") Square in Moscow on May 6, 2012, right before Putin was to be reinaugurated as president, went nowhere; in that instance, the regime changers bit off more than they could swallow, and their local operatives in the "opposition movement" are now some of the most widely despised people in all of Russia.[4] But almost the same technology did work later during the Euromaidan Revolution in Kiev in February of 2014.

When those tasked with protecting what's left of public order become sufficiently worn down to react forcefully when the situation calls for it, the stage is set for...

THE OCCUPATION

During this phase, which, if effective, is quite short, the protesters storm and occupy a symbolically important public building. This is a very traditional revolutionary tactic, going back to the storming of the Bastille in Paris on July 14, 1789 or the storming of the Winter Palace in St. Petersburg on November 7, 1917. If the preparations were successful, by this point the government is too internally conflicted to act, and the defenders of public order are too demoralized to follow their orders, or both. In some cases, as in Serbia, in Georgia and in Kyrgyzstan, this is all it took to move on to the next phase. The highly organized people behind the supposedly spontaneous blitz now declare themselves as the legitimate government and demand that the real government obey them and step down. However, sometimes this doesn't work, in which case there is always...

THE MASSACRE

Mercenaries with sniper rifles are flown in and ushered into the upper floors of public buildings overlooking city squares where rallies and demonstrations are being held. By this time the defenders of public order are sufficiently demoralized by their inaction in the face of increasingly brazen challenges from the protesters

---

4   Hilariously, the little white ribbons, which were shipped into Russia from somewhere just in time for this action, had also been worn by Nazi collaborators during World War II—something many Russians knew while the foreign puppet-masters behind the fake protests clearly didn't.

that a few of them can be easily corrupted by large bribes from the foreign sponsors of the regime change operation. They accept the money and depart the scene, leaving doors unlocked or even handing over the keys. The mercenaries go to work and kill a hundred or so people. At that, the Western-controlled mass media immediately express outrage, pinning the responsibility for the massacre on the government and demanding that it resign. The protesters are incited to immediately echo these slogans, and a groundswell of misdirected outrage sweeps the government out of power, setting the stage for...

REGIME CHANGE
The new government, hand-picked by the US embassy and the US State Department, assumes power and is immediately granted recognition and support by Washington and its allies, along with fawning coverage in the Western-controlled mass media.

• • •

THE COLOR REVOLUTION strategy can be successful—to a point. As we shall see, society can and sometimes does develop effective antibodies against it. It is notable that just about any government—from the most democratic to the most autocratic and authoritarian—is susceptible to it, the only real exceptions being absolute monarchs who can make heads roll the moment someone starts speaking out of turn or those rulers who derive their legitimacy from a divine right that cannot be questioned without committing sacrilege.

The government has no good tactical options. It cannot declare the mass of protesters as outside the law, because they are, after all, its citizens, and most of them are not even directly guilty of any administrative transgressions. Nor can it act against the foreign instigators, many of whom are VIPs or have diplomatic immunity, without provoking an international incident. If it is to restore public order, it must crack down on the demonstrators. If it cracks down early, then it looks heavy-handed and authoritarian, handing ammunition to the protest movement. If it cracks down at the height of the protests, then it causes a lot of unnecessary casualties, turning much of the population against it. And if it attempts to crack down when it is too late, then it only ends up looking even weaker than it is, accelerating its own demise.

But the government does have an excellent strategic option, provided it lays the groundwork for it beforehand. The problem with opposing this sort of "nonviolent" externally driven regime change operation is that it cannot be effectively opposed by a government. But it can be quite effectively opposed and disrupted by a relatively small group of empowered individuals acting directly and autonomously on behalf of the people.

Terrorism by Proxy

We will not discuss this third method of regime change in any great detail because, frankly, it doesn't work. It has yet to result in the installation of a stable puppet regime in any of the countries where it has been tried. It failed in Afghanistan: after the Soviets finally withdraw, the country became a failed state. America's pet terrorists, termed al Qaeda, were then used as decoys to justify invasions of Iraq and Afghanistan, but the decoys came to life and destabilized the region. The latest group of America's pet terrorists, ISIS,[5] who, as of this writing, are so impressed by the Russian bombing campaign against them that they are busy shaving off their beards and running away, has become a huge embarrassment for the US. Terrorism by Proxy does reliably produce failed states, and although some may claim that this is a reasonable foreign policy end-goal, it is very hard to argue that it is in any sense optimal. Moreover, of all the political technologies, Terrorism by Proxy is the most likely to fail through internal contradiction: fighting terrorism while supporting terrorists is bound to eventually

---

5   Variously known as ISIS, ISIL, the Islamic Caliphate, and Daesh.

cause cognitive dissonance even in the most inattentive and
ignorant population.

A requiem

IN A SENSE, this is a requiem for these three political technologies.

The first one—International Loan Sharking—is not going to
work too well going forward. Developing countries can now borrow
from China's Asian Infrastructure Investment Bank, in which they
can become shareholders. Countries around the world are unloading
their dollar reserves and entering into bilateral trade arrangements
that circumvent the dollar system. With its own finances in disar-
ray, the US is no longer able to function as the purveyor of financial
stability. It is now very much a purveyor of financial instability, but
that is not a product that will find too many buyers.

The Color Revolutions have also largely run their course,
because the political technology for neutralizing them is by now
quite advanced. The latest large-scale effort—in the Ukraine in
2014—has resulted in a failed state. Subsequent efforts in Hong
Kong and in Armenia fizzled.

Lastly, Terrorism by Proxy not only never worked correctly, but
is now poised to prove hugely embarrassing for the Washington
establishment. The Russians, with Syrian, Iranian and Iraqi help,
are swiftly rubbing out America's pet terrorists with equanimity
and poise, while their erstwhile supporters in Washington are visibly demoralized and spouting preposterous nonsense. But there
are still some important lessons to be extracted from all this—and
we should extract them before it all gets covered by a thick layer
of dust.

Beneficial uses of political technologies

WHAT WOULD A successful strategy for bringing about rapid
social change look like, such social change being necessary if we are

to avoid the worst ravages of environmental and social destruction and catastrophic climate change? Can we use political technologies to rapidly introduce naturelike technologies and bring the technosphere back into balance with the biosphere?

To recap, political technologies can be used to pursue the following aims:

1. To improve everyone's welfare by pursuing the common good of the entire society. (These are the good kind.)
2. To enrich, empower and protect privileged elites and special interests at the expense of the rest of society. (These are the bad kind.)

Alas, most of those living in the US have had exactly zero exposure to political technologies of the first kind. And so we have delved into methods which the US has been using to undermine countries around the world, because in response the rest of the world has evolved successful political anti-technologies of its own and is in the process of neutralizing the US on the world stage. The US has been able to operate virtually unopposed during the years since the collapse of the USSR, laying waste to entire countries in several regions around the world. The ongoing failure of the US effort to overthrow the government of Syria is significant, but it is not yet enough.

No doubt, it would be exciting and inspiring to expound on the virtues of positive political technologies, but it would also be a pointless distraction. Before we can work for the common good, we must first stop the spread of the common evil.

The importance of patriotic leadership

THE POLITICAL TECHNOLOGIES of International Loan Sharking and of Color Revolution only succeed in cases where the local leadership is corrupt and has sold out to foreign interests. In every case where a Color Revolution has succeeded—overthrowing Shevardnadze in the Revolution of the Roses in Georgia (2003), Kuchma in the Color Revolution in the Ukraine (2004), Akaev in the Tulip Revolution in Kyrgyzstan (2005), Ben Ali in Tunis (2011), Yanukovych in the Ukraine again (2014)—the leader in question was a Western stooge. Such national leaders are led like lambs to the slaughter: the Americans fatten them up and treat them well until the very moment when they are to be done away with. At that crucial moment they are given a choice: run away or be killed. And so they all run away.

The exception that proves the rule is Egypt's Hosni Mubarak in 2011: he refused to run. The Americans had prepared the entire protest scenario. The US embassy in Cairo flew a certain hand-picked youthful leader of the rebellion to training seminars in the States, handed out 26-page protest manuals and set the mayhem in motion with plentiful support from US-owned social media such as Twitter and Facebook, and a chorus of favorable news coverage in Western-controlled mass media. But instead of fleeing, stubborn old Mubarak pulled the police and the army off the streets, retreated to his summer residence in the tourist mecca of Sharm el-Sheikh and simply let Cairo descend into chaos.

Mubarak's strategy eventually failed because Egypt's entire security and defense establishment had been stocked with stooges and traitors. But in general his strategy was correct. Withdrawing allows the protesters to run amok and the protest organizers to lose control. After a short time, seeing that the police are impassive or gone, looters swell the ranks of protesters. This causes the local population to turn against the protesters, organize local self-defense units, construct barricades and so on, and the revolution begins to choke on its own vomit.

This impasse prompts the revolution's organizers to move on to staging a massacre, sending in snipers who indiscriminately kill both police and protesters. Western-controlled mass media then blame the massacre on the government, demanding its resignation, and the organizers get the protesters to echo this demand. This is a difficult problem for the government, because in the midst of protests with many public buildings inaccessible, it is almost impossible to maintain a watertight security environment that would not allow mercenaries with sniper rifles to filter in and take up positions. Nor is it possible for the government to defend itself after the fact, when it has already been discredited, by challenging Western propaganda with propaganda of its own, even if the truth is on its side. After a concerted attempt to sentence Mubarak for all manner of crimes, he was cleared of all charges, but this did not change the outcome.

But what if Mubarak wasn't just a corrupt local oligarch in the pay of Western interests? What if he were a true Egyptian patriot? What if Egyptians trusted him, felt that they shared his values and looked up to him not just as a political leader but a moral and spiritual one as well. Such a leader does not require the heavy-handed tactics of an overbearing security apparatus because the people will come to his defense.

The need for partisans

A SITUATION WHERE the legitimate authorities are politically weak (because of outside pressure) but morally strong and have the truth on their side allows groups of locals to come together and form cells of partisans. While united by a common strategic goal—to defend their communities, thwart the outside forces, uphold legitimate authority—they are completely free to choose their tactics. Because they are spontaneously, anarchically organized, such partisan groups can be far more nimble than the government. Nor do they need to constrain themselves to tactics that are strictly legal. Here are some of the tactics that the partisans can add to their arsenal:

• Use multiple methods, from face-to-face communication with small, local groups at the neighborhood level to the use of social media, to get the truth out: that this is a foreign-organized, foreign-funded campaign based on lies. Detail what these lies are, present the evidence, and let the people draw their own

conclusions. Since these local groups do not pretend to be an official source of information, they are invincible against the charge of
spreading propaganda. The most the foreign puppet-masters can
do is claim that they are "trolls" paid for by the other side—a story
that other locals who have a good sense of who is who are unlikely
to swallow.

- The puppet-masters behind Color Revolutions like to remain
anonymous and "lead from behind," and the goal of the partisans
is to strip them of their anonymity. Suddenly faced with a crypti-
cally hostile, disingenuous "fan club" that monitors their every
movement and picks them out in every crowd demanding a "sel-
fie," making their whereabouts known at all times and generally
hassling them with effusive, faux-friendly familiarity at every turn,
the puppet masters are easily outed and neutralized. By making the
outsiders' identity known, the partisans provide a valuable service
to the local security services, saving them the trouble of spying on
or infiltrating the protest movement.

- Co-opt demonstrations by injecting specific issues and slogans
that resonate with the local population. During Color Revolutions there are usually organizers lurking in the background who
are "leading from behind" by quietly telling people what to shout
based on a pre-approved script. The slogans are generally about
nothing—"freedom" and "democracy" and other such nonsense—
because they can't very well be about the real goal of overthrowing
the legitimate government through nefarious means. By injecting
slogans in pursuit of specific, locally significant demands—"Lower
bus fares!" "Freeze tuitions!"—the partisans can make the protest
be about something legitimate that is potentially a win-win. The
government can then step forward, announce that it has heard the
voice of the people and negotiate in good faith. The protest movement then dissolves in jubilation—"We won!"—the government
takes credit for a successful exercise in direct democracy, and the
puppet-masters go home with nothing.

- Splinter the protest movement by creating a large number of social
organizations. When the Color Revolution organizers try to hold

a meeting, the partisans can try to inject a different agenda, claim that the real venue is elsewhere, show up in numbers and put forward a different leadership, stage a protest against those running the meeting and walk out, taking some number of others with them and so on. If written instructions are handed out, or props such as ribbons and placards, inject different instructions and different props that pursue a legitimate, local agenda.

- Liaise with state security services and local authorities, and trade detailed, real-time intelligence in exchange for specific small favors. Make these favors available to members of the protest movement in exchange for some behavioral changes or compromises. This can often be presented as the work of protest sympathizers within the government to be taken as a sign that it is about to collapse and can bolster the partisans' standing among the protesters.

- Provide the security services with legitimate targets. Much of the work of the Color Revolution organizers involves gradually eroding the boundaries of permissible behavior until anything goes and the security forces, having allowed numerous minor transgressions, have become demoralized and are unable to mobilize against major ones. The organizers try to use human shields in the form of "children"—young, innocent, naïve, chanting about freedom and democracy—who then violate public order in minor ways. "But they're just children!" and so the police do nothing. But if among these "children" there are some partisans who resort to a bit of staged violence here and there, with some pushing and shoving and a few punches thrown, providing the authorities with the excuse they need to intervene, then this pierces the veil of "nonviolence." Remember, blocking streets and hindering public access to public buildings are not, by any stretch of the imagination, nonviolent acts. "Nonviolence" is nothing more than a tactic. It can even be used to promote violence by rendering a population defenseless in the face of aggression, in order to provoke a massacre and then use it for political aims, as was done by Gandhi, who preached nonviolence to Hindus, profiting politically when they were then massacred by Moslems.

- Organize local self-defense units. Patrol neighborhoods to prevent looting. Intervene in demonstrations to keep the protesters in line, helping the security services accomplish things that they may otherwise find difficult to justify. If the government can demonstrate that it's just having a bit of trouble reigning in some patriotic-minded elements within the local population who rose up in spontaneous opposition to the protests, then claims of government heavy-handedness begin to ring rather hollow.

- Out of the stronger self-defense units, organize commando units and train them for special missions. These can be deployed if the Color Revolution proceeds through the stage of massacre and all the way through to regime change. At that point, the hand-picked puppets are about to be ensconced in official buildings, granted official titles, given fawning press coverage by the Western-controlled press and swiftly granted diplomatic recognition by Western governments. But before this can be accomplished, they have to be briefly trotted out before the public to create the illusion that they are "of the people." It is at this point that they are at their most vulnerable, and all the previous efforts to splinter, co-opt and destabilize the protest movement can be brought to fruition to neutralize the would-be puppet government through a few decisive actions. Since by this point the puppets are being guarded by foreign mercenaries who are professionals, the commando units should likewise be composed of people who have professional discipline, training and experience. The installation of a puppet government is a political exercise which, in order to succeed, has to be successfully misrepresented as a popular triumph and as such can be derailed by a public embarrassment or a panic. Also, it helps to remember that the puppets are being installed by mercenaries, who, by their nature, are allergic to the idea of dying, since being dead gets in the way of collecting their pay. If their work environment becomes sufficiently dangerous, they reliably run away.

Finally, if all else fails, the ultimate recourse is an armed uprising based on a guerrilla movement. If the movement has local public support, it can sustain itself for many years. In order to win, a guerrilla movement simply has to persist. After a few years of being unable to control its own territory, the state headed up by the puppets comes to be regarded as a failed state and an embarrassment for the puppet-masters, who are then forced to cut their losses and pretend that the problem doesn't exist. The state can later be resurrected minus the puppets, or fission into several smaller statelets. By the way, this is precisely what is happening in the Ukraine as I write this: the armed uprising in the east (the industrialized, educated, Russian-speaking, densely populated part of the territory) has left the central authority in Kiev circling around in an ever-expanding void, unable to either crush the rebellion in the east or to accede to internationally agreed-upon terms for granting that region autonomy, since this would undermine its raison d'être of building a monolithic ethnically pure Ukrainian state. As the void deepens, it is becoming an ever-greater embarrassment to its masters in Washington.

The making of a partisan

IF A PARTISAN movement can rise up out of nowhere to defend local communities against foreign usurpers, then can a similar partisan movement form to defend the local biosphere against the depredations of the technosphere perpetrated by its servants in international finance, transnational corporations and national governments that have surrendered their sovereignty to Washington or to its proxies in Brussels, at the IMF and the World Bank? I would like to believe that this is possible.

But what makes a person a partisan? What made the coal miners, factory workers and taxi drivers of eastern Ukraine take up arms in defense of their communities? There are many ingredients, but the following few seem more important than any of the others:

• A separate sense of cultural, religious and linguistic identity. The Ukraine's eastern provinces originally formed the Tauric Governorate of Russia and remained part of Russia until quite some time after the Russian Revolution of 1917, when Vladimir Lenin reassigned them to the Ukrainian Soviet Socialist Republic in an effort to "civilize" it and make this backward, agricultural, historically foreign-dominated region ready to accept Soviet socialism. The people there consider themselves Russian not Ukrainian. They speak Russian; they are culturally Russian; and they are Russian Orthodox, unlike the Uniates of western Ukraine who were Orthodox once but have been converted to Catholicism.

- A keen sense of history. Eastern Ukraine was the scene of some of the most pitched battles with Nazi Germany, and so when the Ukraine's new puppet government made heroes of Nazi collaborators and when people from the west of Ukraine marched with torches and Nazi insignia, the people in the east were naturally inclined to think that they needed to finish the work of their forefathers and defeat Nazism once and for all.

- A strong perception that the new central government is illegitimate. It was installed in the wake of a violent overthrow of the legitimate, constitutional government which was headed up by Viktor Yanukovich who, for all of his many failings, enjoyed a great deal of support in the east of the country. It did not help matters that the membership of the new government was hand-picked by Washington and that it included foreigners. In this, they saw the need to defend their native land against foreign invaders.

- A sense that, for all of the limitations of their support, their Russian cousins across the border have their backs. Although NATO talking heads endlessly waxed hysterical about "Russian aggression" and "Russian invasion," no evidence of it exists. But what does exist are volunteers and weapons filtering in across the purely notional border, lots of humanitarian aid convoys, help with intelligence from the Russian military and a Russian government willing to put pressure on the one in Kiev to grant autonomy to its eastern provinces (enshrined in international agreements Minsk-I and Minsk-II, which the Kiev government has so far refused to fulfill). The people in the east were left hoping that Russia would do more, but Russia preferred to act in accordance with international law and work through international institutions to resolve the conflict.

- A sense of victory. The Ukrainians attacked the east not once, not twice but three times. Each time they were humiliated, suffering horrific losses. Now the most they can do is sporadically lob missiles into civilian districts, causing a lot of death and injury but failing to advance Kiev's or Washington's agenda a single millimeter. (Remember, Terrorism by Proxy doesn't work.) Had the insurgency suffered a defeat at the hands of the much larger and much better-equipped Ukrainian military, it would probably have melted away along with the partisan movement that gave rise to it.

Partisans of the biosphere?

TO ABSTRACT AND summarize the ingredients of a successful partisan movement, the following elements seem to be key for motivating people to join and for making the movement a success:

1. A uniting **ideology**, a common set of cultural markers or a common code of behavior that imbues the participants with a sense of purpose, allows them to recognize each other and enables them to work together toward a common goal
2. A **groundedness** in the environment, which is not an abstraction but their local, specific habitat, giving them a sense of belonging and a sense of responsibility for defending and protecting it
3. A sense of **outrage** at being despoiled and violated by an enemy that must be held at bay and eventually vanquished because their very survival is at stake
4. A sense of **belonging** to a larger whole whose members, even if they cannot render immediate assistance, do provide moral support and a sense of legitimacy and common purpose

What would these key elements be for a partisan movement that could dethrone the technosphere?

The ideology might take this very book as its source, the

common set of cultural markers may naturally evolve from trying to apply it in daily life, and the common code of behavior can be based on the imperative of ratcheting down on the harm/benefit hierarchy, avoiding technologies that have unlimited harm potential and moving toward technologies that are naturelike and zero-harm.

The groundedness in the local environment can only come from a deep and abiding attachment to nature—not the biosphere as an abstraction, but specific forests, meadows, streams, rivers, coastlines and the plants and animals that inhabit them. It comes from the conception of wild nature as that which is most sacred and the source of all spirit, human spirit included. In a religious context, it is the belief that there is no god beyond our one and only living planet[6] and that the story of humans is either part of the story of that living planet, or it is a fake story of human uniqueness and superiority concocted to make us submit to authority and to make us serve the technosphere.

The sense of outrage can come from the visceral sense of the amount of damage that has already been inflicted, amplified by the intellectual understanding that the biosphere is in extremis. It can come from walking down a once-pristine beach and finding it covered with plastic trash or from revisiting a favorite childhood fishing spot or frog pond and finding it contaminated by an oil slick or filled with worn-out tires. It can come from going for a walk in a forest and finding a once beautiful grove reduced to stumps and slash piles. A lot of people see such things and feel outrage tinged with grief, but that is not the reaction of a partisan. The

---

6   It may be helpful to appreciate that the vast, infinitely remote, impersonal horror that is the universe, with its billions of galaxies infested by supermassive black holes, is nothing more than a scientific freak show which, in the best possible case, will have absolutely no bearing on our fate. Of course, we can be grateful for having something to marvel at and that it hasn't killed us yet with an asteroid strike or a burst of sterilizing cosmic radiation.

proper response is a steely resolve and a cold, calculating approach, informed by an understanding of the technosphere's nonhuman, machine-like nature. What is needed in order to defeat it is not emotionalism but a method composed of strategy and tactics. The strategy is described in this book; the specific tactics are entirely up to you.

The sense of belonging may take time to achieve, but as the movement grows and spreads to diverse localities and lands, there may emerge a sense of kinship among all of those who embark on this journey, strengthened by the realization that the many disparate, local efforts are gradually combining into a single whole.

Finally, the sense of victory can come from individual action, starting from the very first step taken down the harm/benefit hierarchy and developing over time as individual actions inspire group actions, which open up more and more possibilities for autonomy, self-sufficiency and freedom, in turn causing the technosphere to visibly, perceptibly shrink.

7
=

SOCIAL MACHINES

ALTHOUGH THE TECHNOSPHERE gains most of its power from the use of technology—specifically, the burning of fossil fuels and the mining and refining of nonrenewable natural resources, using them to manufacture short-lived products, many of which are environmentally harmful or of dubious value—surprisingly, its most important moving parts are not machines but people. More specifically, it is people who behave like machines: people who have sacrificed their autonomy and suspended the use of their own judgment for the sake of a steady paycheck (or a reliable return on investment) and a sense of security (which may very well turn out to be false).

Although the technosphere exerts its power and control over individuals, it gains that power and exerts that control not through individuals but through specifically organized groups: **social machines**. A social machine is a form of organization that subordinates the will of the participants to an explicit, written set of rules, that is controlled based on objective, measurable criteria, and that excludes, to the largest extent possible, individual judgment, intuition and independent, spontaneous action. In the process, it becomes blind to all the things that cannot be measured, such as meaning, beauty, happiness, justice and compassion.

The progression from a humanistic organization that functions on the basis of common understanding, spontaneous cooperation, shared values and individual judgment and initiative to a social machine in which people behave like robots, is automatic: it is simply a question of scale. As a family business or a local club expands to the point at which it becomes a multinational corporation or an international nongovernmental organization, it becomes necessary for it to impose a formal management structure and to require that its members strictly adhere to an increasingly vast and detailed set of rules and procedures. The impetus for long-term sustainability and responsiveness to the needs of its members is gradually replaced with the need for growth and dominance within a competitive global environment. The need to compete displaces cooperation throughout the organization, and what may once have been a mutually supportive, caring, nurturing environment is replaced with the alienation, cold professionalism and personal indifference of the modern office.

Once every expression of humanity has been purged from the system and replaced with a set of bureaucratic functions and rule-following behaviors, an organization becomes what sociologist and historian Lewis Mumford called a social machine. It is inherently inhumane, and even though staffed by humans, it is in its functioning indistinguishable from what we would expect from a robot.[1]

One key feature of social machines is that they do not actually exist: they are mere figments of our collective imagination. For example, the us government is said to exist only by virtue of the fact that millions of people *believe* that it exists and behave accordingly as a matter of self-preservation. If one day they were to become convinced that it no longer exists—because of space aliens or, more likely, because it has become too internally conflicted and

---

[1]   Many modern social machines are increasingly automated, with most of their human components replaced by internet servers and robots, serviced by a handful of technicians who decide nothing.

impotent to matter—they would stop showing up for work in government agencies, they would disregard federal laws, and they would spontaneously start organizing alternative power structures. And if they did this, then the US government would indeed cease to exist. When a social machine fails it disappears without a trace. Usually it leaves behind some vacant office buildings, but a building is merely its empty shell.

Thus, social machines are not really physical entities but modes of behavior. If a machine is typically described as a "method and apparatus"—the phrase found on patents—a social machine is a method for interacting with people, and the apparatus is a group of people and some office equipment. Being part of a social machine is a mere spell, and a spell can sometimes be broken.

Part-human, part-machine

THE ABILITY TO control large groups of humans so effectively that they become indistinguishable from robots developed quite recently, and here technology—specifically, information technology—has played a key enabling role. The printing press made it possible to promulgate vastly more laws, rules, procedures and instructions. The typewriter, the adding machine and various tabulation machines made it possible to collect reports and to distill statistical information from them, providing objective criteria for making decisions. Later, the advent of computers made it possible to automate much of this decision-making, removing the very possibility of exercising human judgment. The telegraph, the telephone and the internet provided secure channels for exerting real-time control over geographically distributed operations. High-speed internet has made it possible to respond within a fraction of a second—an ability that is exploited by high-frequency traders in the financial markets (not the traders themselves, mind you, but their automated trading algorithms).

But it wasn't so long ago that even armies, which are the most disciplined and regimented of organizations, had to leave much of the decision-making to the personal initiative of individual officers. When sealed orders had to be delivered on horseback, they were inevitably out of date by the time they reached the units on the battlefield. Best-laid battle plans often became obsolete on first contact with the enemy, and the fog of war provided little scope for collecting real-time information to serve as a basis for objective decision-making. The application of rational management methods to military operations was further complicated by pervasive nepotism within the officer class: high-ranking officers could hardly be expected to be objective when issuing orders to their cousins or nephews. Lastly, morale critically depended on the men's trust in their officers, which was in turn determined by the officers' ability to keep the men alive and to provide them with food, drink, women and booty. This made officers rather difficult to reassign even if they showed clear signs of excessive personal initiative or insubordination.

All of that has changed, thanks to modern gadgetry and communications technology. The modern soldier is part-human, part-machine: wired for sight and sound, located and guided by satellite signals, always and everywhere painted as a spot on the map in some situation display, his every act and motion monitored and recorded. In this environment, officers play it safe and try to do everything strictly "by the book" because even a slight deviation may negatively affect their career. Since so little depends on the individual officers, the men become indifferent to who leads them; in turn, this makes it easy to rotate officers in and out without affecting morale. The officers begin to see the men, and the missions, as mere stepping stones to the next promotion or transfer. Their adversaries, who are, more and more often, non-state actors, love this state of affairs; after all, the best kind of enemy to have is a perfectly predictable one.

Similar conditions prevail in the modern office environment, which is virtually a panopticon. Everyone is at all times trackable, via the laptop, the smartphone and the ever-present security cameras. They clear in and out of secured areas using security cards or key fobs. The latest trend is to implant employees with RFID² tags, like pets or cattle, to save them the trouble of carrying around an object. They remain trackable even when they are outside of work, because it is now a common requirement to be able to respond to phone calls and e-mails around the clock, even when taking vacations (which have been re-dubbed "workations").

In response to having their every step and every word monitored and recorded, people try to play it safe: follow the rules, stick to the script, and not ask too many questions. This, they think, is the best way to make sure that they get to keep their job. But the truth of the matter happens to be somewhat different: by agreeing to act like robots, they make themselves easy to *replace* with robots. After all, a robot can read from a script or follow a step-by-step procedure just as well, if not better, than a human.

A playground for psychopaths

PSYCHOPATHS—INDIVIDUALS WHO HAVE no empathy or moral sense and are forced to simulate them in order to function in society—normally make up a small percentage of the general population. In a healthy society they are shunted to the margins and sometimes shunned or banished altogether. Sometimes they can take on an interesting, marginal role for which total lack of empathy or conscience is a boon: executioner, assassin, spy... In an environment where people take care of each other—because they feel empathy for one another—psychopaths stick out like a sore thumb. Even if they can simulate sincere expressions of empathy to some limited extent, they usually can't fake them well enough to keep people around them from growing apprehensive, and just one or two episodes that demonstrate their indifference to others' suffering or a sadistic streak is usually enough to "out" them conclusively.

But what to a healthy society looks like a terrible character flaw appears perfectly normal, even laudable, in the context of

2   Radio-frequency identification

a social machine. Lack of empathy is seen as cool, professional detachment; a psychopath would never let emotion cloud her judgment. Sadistic tendencies (psychopaths hurt people in order to make themselves feel *something*) are perceived as signs of an incorruptible nature: the rules are the rules! Conversely, while a normal person feels alienation when thrust into an alienating environment, finds it painful to act like a robot and suffers pangs of conscience when forced to inflict damage on others by blindly following inhumane rules, a psychopath feels nothing at all. Because of this, social machines act as psychopath incubators. Psychopaths are not the healthiest of specimens, but because of their greater *inclusive fitness* within social machines, psychopaths tend to persist and thrive within them while non-psychopaths do not.

In turn, in societies dominated by social machines, one's ability to thrive within a social machine is a major determinant of one's ability to create positive outcomes for oneself and one's progeny. Simply put, in such societies psychopaths do better socially, and are therefore more likely to breed successfully. And since, based on research on twins, psychopathy is roughly half-genetic and half-environmental,[3] societies dominated by social machines selectively breed psychopaths. This, in turn, provides more human raw material for social machines, allowing them to grow and proliferate. After some number of generations of such selective breeding, society passes a threshold beyond which it becomes unable to return to health even once its social machines collapse (as they all do, eventually) until enough of the psychopaths have been winnowed from the gene pool—a process that can likewise require a few generations.

If having some psychopathic tendencies is helpful for fitting in within a social machine, having more psychopathic tendencies

---

3   Larsson, Andershed, & Lichtenstein, "A genetic factor explains most of the
     variation in the psychopathic personality." *Journal of Abnormal Psychology*, 115
     (May 2006).

is even more helpful. Consequently, within social machines, pure psychopaths rise through the ranks and concentrate at the top. It should be entirely unsurprising, then, that when we look at the upper echelons of business and government—the C-suite, the boards of directors, the executive branches, the legislatures and the courts—we find that they are pretty much stocked with total psychopaths. This being the case, it seems rather clueless for anyone to think that a society that has been dominated, and sickened, by social machines over many generations can somehow be nursed back to health by its selectively bred psychopathic leaders. These leaders are the symptoms of the disease, and symptoms have never cured anyone of anything.

It's robopaths all the way down

IF PSYCHOPATHS ARE uniquely well adapted to life within a social machine, the rest of us are anything but. Unlike the high-functioning psychopaths that leverage their lack of normal human emotional response and float to the top in the elaborate hierarchies of which social machines are composed, relatively normal people tend to settle to the bottom. In the process, they tend to suffer psychological damage.

When people are thrust into a controlling, despotic context and forced to act like robots, their inner emotional life becomes unhitched from the dispassionate, professional public image they are required to cultivate. The resulting state of mind has been given various names in the psychiatric literature, such as *radical estrangement* and *hyperalienation*. These also happen to be symptomatic of schizophrenia, but this is not schizophrenia: schizophrenics are not renowned for their ability to abide by social and cultural norms and to follow elaborate systems of bureaucratic rules, while these people, for as long as they are able to hold on to their jobs, manage to do just that.

Sociologist Lewis Yablonsky referred to such people as **robopaths**:

People may in a subtle fashion become robot-like in their interaction and become human robots or robopaths. This more insidious conclusion to the present course of action would be the silent disappearance of human interaction. In another kind of death, social death, people would be oppressively locked into robot-like interaction... In this context, the apocalypse would come in the form of people mouthing ahuman, regimented platitudes on a meaningless dead stage.[4]

Robopaths are those who, in being compelled to follow and enforce arbitrary rules, have lost the normal ability to identify with the rest of humanity and have found a new sense of personal identity in becoming sticklers. Because their newfound identity is artificial and fragile, they see any sign of nonconformity and any break with the official norms and rules, not just as a transgression but as a personal affront. In imposing harsh, unjust punishments while following "the letter of the law," they insist that it is "nothing personal," because if it were, it would threaten their sense of who they are. Robopaths cannot allow themselves to act human because that way, for them, lies schizophrenia.

Being forced to act like a robot is a form of oppression, and it inevitably gives rise to pent-up rage and, eventually, to a violent response. But a robopath (with the exception of the policeman and the soldier) cannot be overtly, physically violent, for that would violate the all-important rules.[5] As Sean Kerrigan wrote in his book *Bureaucratic Insanity*, robopathic violence

... manifests itself in cryptic, socially acceptable ways—when a school administrator harshly punishes a student for a minor transgression, or when a bureaucrat takes a perverse pleasure in making it illegal to feed the homeless. It is in such moments that

4   Yablonsky's most famous book is *Robopaths* (Pelican, 1971).

5   The level of robopathic infestation within a society is reflected in its level of police violence and use of excessive force.

the especially venomous nature of bureaucratic violence can be observed. The rule enforcer—simultaneously a mild-mannered and congenial professional and an aggressive and unsympathetic monster—is unable to act out repressed rage in any other socially acceptable way than by doling out punishments, fines, rejections, expulsions and other forms of objective, systemic violence.[6]

It would be bad enough if, in being forced to deal with social machines, we were dealing with machines because at least we would know what to expect. But, worse yet, they are *broken* machines: their top echelons are full of psychopaths who take perverse pleasure in torturing us, while their lower tiers are stocked with knuckle-dragging robopaths who will hurt you just to make themselves feel better, all the while rationalizing that they are doing so because "the rules are the rules." As society decays, the social machines within it degenerate, descending further and further into bureaucratic insanity. When the day comes that the rules can no longer be followed at all, the psychopaths start running amok by making up their own rules as they go along, while, in response, the robopaths suffer depersonalization and anomie and go well and truly insane: an explosive combination.

Countermeasures

WHEN LIVING WITHIN a society that is dominated by the technosphere, we have no choice but to deal with social machines. Here, it is important to view social machines as projections of the technosphere and, as such, a form of technology. When looking for ways to handle social machines that will minimize the damage they inflict on us, we are forced to invent a system of countermeasures: an anti-technology. As with most forms of technology, this will cause some amount of harm while providing certain benefits. The greatest benefit of such an anti-technology is, clearly, in the harm

---

6   Club Orlov Press, 2016, http://www.amazon.com/dp/1530989523

it inflicts on social machines, while its harm to us is mainly spiritual and psychological. In finding successful ways to interact with social machines, we must integrate with them, and this, in turn, requires us to learn to act like bureaucrats. Obviously, it is preferable for us to learn to impersonate high-functioning sociopaths than low-functioning robopaths, but neither choice is psychologically or spiritually healthy.

Not all psychologically healthy humans are capable of putting their humanity on the shelf for the sake of waging battle with a social machine. How many people do you know who went temporarily insane while their court case or application or proposal was pending before some inscrutable bureaucratic body, constantly rehashing its details and unable to concentrate on anything else? What's worse, have you noticed that, just as this is happening, they are unable to judge the full extent of the damage this is causing to their psyche? Such situations often call for an intervention. It is very helpful for a community to have one or two designated social machine wranglers—people who can live by the old adage "Sticks and stones will break my bones, but words will never harm me" without suffering any adverse effects. Perhaps that puts them somewhere on the psychopathic spectrum, but so what? As I already said, psychopaths can be very useful in certain specialty roles: executioner, assassin, spy... social machine wrangler.

Sometimes life thrusts us into a situation that allows us to assess the damage: to what extent have we been touched by bureaucratic madness. For example, suppose you find yourself in the waiting room of some minor local official. The door to the office is closed, there is no receptionist, and there are plenty of people milling about. Periodically the door opens and people walk in or out or stick their head in to ask a question. There is no semblance of a queue, and people are constantly walking in and out of the waiting room and conversing privately. You inquire as to who is last in line, and people look around, shrug and tell you that "he isn't here right now."

There is a range of possible reactions to such a situation. One is to feel a sense of wonder at the functioning of a system you don't understand and start figuring it out by striking up friendly, curious conversations with people, trying to puzzle out how you could help them so that they would want to help you to navigate this strange environment. Another is to feel rage at such a stupid, disorganized, inefficient way to run things: of course there should be a queue, ideally using one of those things that dispense slips with numbers... If you feel wonder, then you are probably unaffected; if you are close to rage, then you have been touched by robo-madness and are in serious need of deprogramming.

But back to battling social machines: lacking access to a specialist who could wage battle on your behalf, your best bet for avoiding psychological damage is to avoid social machines altogether by doing whatever you can to stay off their radar. The first step for any social machine is to classify you, and if you make yourself hard to classify, then you may escape unscathed.

Many social machines are geared to deal with local residents; if you declare that you are a visitor or a tourist, they leave you alone. If you are a resident, then social machines tend to pursue you where you live, and they may determine that the one viable alternative—to live without a permanent, land-based residence—amounts to homelessness or vagrancy and is hence objectionable or even actionable. But if you simply "travel a lot for work" and declare yourself a "business person," then they have little choice but to ignore you.

If you draw a regular paycheck, then this makes you easy to classify and thus a prime target. Even if you avoid a regular paycheck by operating a number of small, unofficial businesses, none of them big enough to be worth going after, then they might consider you as "self-employed" and go after you. But if you are "a person of independent means," even though they are meagre ones, then they have little choice but to leave you alone.

If you have been a very silly boy or girl and managed to get yourself identified, classified and in the crosshairs of a social machine, it is very important to understand that you are dealing with a machine and that no amount of idealism, calls for fairness or justice, appeals to common sense or anything else that isn't part of the machine's programming is going to be of help to you. If you try this approach with a psychopath, you'll just get a laugh; but there is a difference if you try it on a robopath—you are likely to trigger a violent outbreak (expressed in a cryptic, veiled, bureaucratic form). On the other hand, it is sometimes possible to extract some small amount of preferential treatment from a robopath—by convincing one that you are "one of them" by showing respect and enthusiasm for following their rules. However, this tactic is useless with a psychopath, because for a psychopath there is no "them." But there is no harm in trying (provided your sense of personal dignity is sufficiently bulletproof).

But in general it is a waste of time to approach a machine as if it were a human. Instead, you should treat the entire interaction as a purely technical matter. You need to understand the functioning of the machine in purely abstract, technical terms, disregarding the machine's own stated purpose or mission and regarding it as a mechanism that can be tricked into doing what you need it to do, which is ideally to simply ignore you. Yes, social machines can be "hacked." Your task is to identify the spot to throw in a monkey wrench and to do it in such a way that your action isn't understood to be an act of sabotage or, better still, isn't even noticed.

One last point: social machines occupy a spectrum. Some are more social than machine, others more machine than social. It is sometimes hard to determine which is which, but in doing so the following dictum is invaluable:

Any rule that cannot be bent must be broken.

If a social machine follows this dictum, then it is a machine in name only: the rules are kept on file just in case someone asks, but human judgment is allowed to prevail most of the time. You can always tell that this is the case if friendly, compassionate people will tell you who will look the other way under what circumstances, who provides special favors for extra pay, which clerk will be happy to misplace your paperwork or pull it out and place it on top of the pile, in exchange for a bit of gratitude, and what you can be sure to get away with, when, where and how. You may become sensitized to the fact that the social machine is a bit of a charade: officials do the absolute minimum to justify their meagre pay, and so once in a while one of them might apprehend you, give you a stern talking-to and maybe even attempt to write you up, but find that his ball point pen is out of ink or that his notepad doesn't have any empty pages left, and so he lets you go.

Another good way to find out whether a social machine is an inhuman monstrosity or the mere appearance of one is to see how far it chases people. If it never gives up and pursues people to the ends of the earth, willing to escalate endlessly and squander countless resources on "getting their man," then it's a machine. But if deciding how far to chase you—for one block or two—before giving up is left up to human judgment, then it is a machine in name only.

As society degenerates, social machines degenerate with it, and in spite of all the efforts at surveillance and automation, people find ways to survive. And if this requires throwing some monkey wrenches into the works, then more and more people will start doing just that. At some point it will become evident to all that most of the social machines have become so degraded that they are mere relics—empty shells maintained for the sake of appearances—while all of the decisions are made outside of them by actual humans applying their individual judgment to situations to which no written rules need apply.

8
=

WRESTING CONTROL

WRESTING CONTROL FROM the technosphere one step at a time, by ratcheting your way down the harm/benefit hierarchy, is the only possibility open to many people, especially those who are older and are encumbered with family and other responsibilities. But those who are younger and freer may not be happy with such a gradual plan of action. They want their freedom, and they want it now! There are forces that prevent them from making such a leap, but are they insurmountable?

The largest difficulty most people face with making a quick transition away from dependence on the technosphere and toward autonomy and self-sufficiency is that this is virtually impossible to do à la carte—by picking and choosing which of the technosphere's many grasping tentacles you will liberate yourself from today, while disentangling yourself from all of them at once amounts to a frightening leap into the unknown. As long as you remain dependent on any of its major services, you tend to remain dependent on all of them and have to pay for all of them, preventing you from saving up the money you would need to extricate yourself. But, as always, there are a few shortcuts you can take and some clever tricks you can try to use.

The iron triangle

FOR MOST PEOPLE making enough money to stay afloat within
the technosphere means holding on to a job. That job has to be
accessible from where you live, and this, for most people, implies
living within driving distance from it or close to public transportation. In most of North America, this means owning a car. In
addition to the job and the car, you also need a place to live. This,
then, is the iron triangle into which most people remain locked. It
consists of a house (because you need a place to live), a job (because
you need to pay for the house) and a car (because you need to be
able to commute to the job).

The iron triangle is a well-designed trap, and escaping from it is
very difficult. It is difficult even for a young, resourceful individual
who can sustain some amount of hardship and doesn't have much
to lose.[1] It is even more difficult for families with children. This is
because it is not possible to get rid of any one vertex of the iron triangle without jeopardizing the others:

- Lose the job—and how will you continue paying for the house and
the car?
- Lose the car—and how will you get yourself to the job? And if you
can't earn money, then how will you keep paying for the house?
- Lose the house—and where will you recuperate between your
work shifts so that you can continue working the job and paying
for the car?

You could get rid of all three vertices of the triangle at the
same time—but then what will you do? To execute any sort of
plan that will make you autonomous, self-sufficient and free, you

---

[1]   There is the inspiring example of the Google employee who saved 90 percent
of his income by living in a truck in the company parking lot: http://www.
businessinsider.com/google-employee-lives-in-truck-in-parking-lot-2015-10.

need savings. But coming up with savings in any of the usual ways (investments, money in the bank) is rather difficult for a number of separate reasons, plus one general reason: the entire system is rigged to prevent you from saving.

Investments are now purely speculative because non-risky investments such as bonds now have negative yields, causing savings to melt away over time.[2] Banks pay effectively negative rates on deposits and can confiscate your savings, which are now viewed as if they were unsecured credit you extended to the bank. In case of a financial crisis your money can simply vanish, as has happened numerous times in numerous places.

Saving is also difficult because of the technosphere's concerted efforts to confiscate your entire disposable income. In the more prosperous areas—the ones where there are jobs good enough to allow you to pay down your debts and still have some hopes of accumulating enough savings to launch any sort of serious escape plan—costs of all kinds tend to rise until just about everyone's ability to save is extinguished, while outside of these few remaining prosperous areas jobs are too scarce and don't pay enough for you to save.

There is also in the US an entire separate strategy of enslavement operating in parallel with the iron triangle, based on debt serfdom. Here, student loans figure most prominently. Federally guaranteed student loans cannot be discharged through bankruptcy. However, there is a trick: start making a lot less money. Then your payments will be based on your income, not on what you owe, and the remaining balance of the loans will be discharged after 20 years. But then, if you aren't making much money, how can you save any?

Then there is operating in the US a financial scam known as private health insurance. This too shows a marked tendency for

---

2   About the only reason why negative-yield bonds are currently worth anything is because they too can be used as speculative investments—up until the moment when the bond bubble bursts.

insurance premiums to increase to a point where almost everyone's discretionary income has been soaked up. But thanks to the hilariously misnamed Affordable Care Act there is a trick here as well, the same one: start making a lot less money. If your income puts you below the poverty line, then your insurance is free.

Lastly, some people have very high credit card balances at high interest rates. These can be discharged through bankruptcy. For some, coming up with the savings to execute your plan for escaping from the iron triangle then involves going bankrupt, then saving up, then letting your official income drop to below the poverty line, then being imaginative while flying below the officials' radar. Or some combination of the above, in some sequence. You have to use your imagination. If some of the things you imagine strike you as illegal, then you have to remember this: in the United States, nothing is illegal until you get caught. Believe it or not, that is the actual law of the land, and those at the top of the economic food chain have absolutely no qualms about using it to their fullest advantage.

There are ways of breaking out of the iron triangle. None of them is particularly easy or obvious—you wouldn't expect the technosphere to make it easy for you to break away from it, now, would you? Quite the opposite: if a certain trick turns out to work particularly well and enough people find out about it and start making use of it, then, chances are, people will start getting caught for trying it, and it will stop working. You have to be sufficiently smart, resourceful, fearless, determined and stealthy, but if you are then there is nothing to stop you.

Distracting ourselves

IT IS QUITE difficult to get much of anything accomplished if you are constantly being distracted. No matter what it is you set your mind to, you won't get far unless you stay focused, and being distracted is the opposite of being focused. But most people go through their daily lives being constantly distracted by images, words and activities that are specifically designed to ruin their concentration. What's more, many people absolutely need to be distracted on a regular basis in order to avoid going insane. Conditioned by life in a world of nonstop background music, omnipresent television screens and round-the-clock news feeds, they go into severe withdrawal the moment all of this artificial stimulation is taken away.[3]

Taking a step back, consider what your life would be like if it were perfect. You'd only engage in activities you found useful or pleasant, preferably both, and only for as long as you wanted. Each day would be yours to decide what to do and how to do it. You would happily isolate yourself from anyone you found unpleasant, tiresome or simply unnecessary to your life and surround yourself with your loving, supportive extended family and a circle of close, true friends. You would have no need to do anything special to maintain appearances and would always be able to act in accordance with your true nature. You would look upon your past life with a deep sense of satisfaction and upon your future with anticipation of even greater satisfaction, and just a bit of wonder to keep things interesting. You would solve problems as they surfaced while not caring a whit about anything that did not concern you. At the end of each day you would feel tired—from having accomplished the few essential things you set out to accomplish that day—and simply relax. Perhaps you'd watch the sunset refracted in ripples on the water, or watch children play, or meditate. Would you feel the need to constantly distract yourself from such a life? I don't think so. In fact, it would be very difficult for anyone to distract you from it.

Now consider what your life is actually like. Do you have to rush about from place to place, constantly dealing with complete

---

3   So severe has this problem become that researchers have found that meditation is now driving people into depression: http://oxfordmindfulness.org/wp-content/uploads/jccp-paper-021213.pdf.

strangers and people you know but don't necessarily like? Is the time you can allot to your family and friends woefully short—because everyone else is also too busy rushing about from place to place and dealing with complete strangers and people they don't necessarily like? Do you have to obey a fixed schedule, execute arbitrary tasks that others have assigned to you and follow rules you had no role in establishing? Do you have to maintain certain appearances for the sake of pleasing strangers and people you don't like? Are you being made to feel responsible for things over which you have no control? Does all of this cause a great deal of stress? And to alleviate this stress and to make your life tolerable, do you find the need to constantly distract yourself—with idle gossip, antics of cats on the internet, professional team sports, political mudslinging, daydreams, alcohol and drugs? If so, then you may find yourself unable to maintain the level of focus that would be necessary to transform your life and bring it closer to the ideal sketched out above.

If you do want to transform your life, where do you start? Well, it turns out that the work of fighting distractions is almost as distracting as the distractions themselves but not quite—because it requires focus. Initially, you don't even have to stop distracting yourself entirely. For example, suppose you watch television (which is a distraction) but you only watch one show, which happens to be quite good, so you are already only focusing on what happens to be good on television. But the show is on a commercial channel which shows ads, which are distractions within a distraction. This gives you a chance to focus on not watching any of the ads—by turning the sound off and working on a sudoku puzzle or a crossword while they play. Distracting yourself from an unwanted distraction is already a victory of sorts.

This is all child's play, but it gets serious when you start exercising your newfound powers of focus to eliminate from your life as many as possible of the things that cause you stress, which, in turn, force you to seek ways of distracting yourself. A reasonable way to start is by setting up a few rules. For example, how many times a day do you deal with your e-mail? If you read and possibly respond to each e-mail as soon as it comes in, then you are probably causing yourself unnecessary stress. But you can make a simple rule: you will only read and respond to e-mails three times a day—in the morning, when you start your workday and are rested and, one would hope, in a good mood; right after lunch, when you are, again, refreshed and restored; and right before you quit for the day, when you are again in a good mood because you are so looking forward to being done with work. Suddenly e-mail no longer has the ability to cause you stress. This too might seem trivial, but it gets better.

One of the greatest distractions of all is shopping. What's more, shopping has a tendency to create a vicious cycle: the more you shop, the more money you have to earn, the more stressful your life becomes and the more you need to distract yourself—by shopping. Here, again, you can unwind that vicious cycle by focusing on what you absolutely need or really, really want.

Make some more rules for yourself. Do you have too many possessions, including ones you hardly ever use? Here is a good rule: if you haven't used something in over a year, get rid of it. Here is another very good rule: for each new possession you decide to add, choose two possessions to get rid of. After some amount of time, when you have decided that your set of possessions has been optimized, you can start getting rid of just one possession whenever you add a new one.

If you do this, your need to make money will be greatly reduced. Instead of focusing on making money while trying to distract yourself from the fact that focusing on making money is causing you stress, you could start focusing on eliminating all the things that cause you stress. Keep this up, and you just might achieve your ideal.

Part of the problem is deciding what is a distraction and what isn't. For example, take electoral politics in the US: is paying attention to politicians in order to decide how to vote an important part of being a citizen, or is it just a distraction? A 2014 Princeton University study by sociologists Martin Gilens and Benjamin I. Page[4] argued persuasively that the US is not a democracy: their statistical analysis showed that in the US public policy decisions are not correlated with preferences of the electorate; instead, they are correlated with the preferences of a tiny part of the electorate composed of business lobbies and the very rich.

This study provides an objective standard by which to determine whether paying attention to electoral politics in the US and spending time deciding how to vote is or is not a distraction in your specific case. All you need to do is ask yourself a question: "Am I a multimillionaire or a business tycoon?" If the answer is "Yes," then, by all means, do pay attention. If the answer is "No," then your participation is guaranteed to be utterly inconsequential and is just another distraction.[5] Once you take these facts on board, blocking out electoral politics may not be too difficult. You may still have some difficulty with all of the people who insist on distracting themselves with it, perhaps yourself included, despite the fact that you are neither multimillionaires nor business tycoons.

Or take climate change: is paying attention to the topic of catastrophic climate change a legitimate focus or a distraction? Here, the situation is a bit more complicated. If you want your children and grandchildren to have a survivable future, then climate change is a legitimate concern, because it won't do well to commit them

---

4   "Testing Theories of American Politics: Elites, Interest Groups, and Average Citizens." https://scholar.princeton.edu/sites/default/files/mgilens/files/gilens_and_page_2014_-testing_theories_of_american_politics.doc.pdf.

5   There is, however, an exception: if you treat voting as a game of strategy, then you can deprive the moneyed interests of achieving the results they pay for by voting randomly (by flipping a coin). You would remain politically powerless, but you would at least be pursuing the long-term goal of discrediting this faux-democratic system, opening up the possibility of creating something better.

to living in a place that will end up parched dry, or too hot and humid to avoid heatstroke, or submerged under the rising oceans, or regularly devastated by monster superstorms. You can perhaps avoid such negative outcomes by paying close attention to climate scientists and by making certain strategic decisions based on their predictions.

But there are other, less useful uses of climate science. One is to titillate yourself with it, as with any other sort of disaster porn, be it asteroids, volcanic eruptions or zombie invasions; this is definitely just a distraction. The other way to distract yourself with it is by paying attention to climate politics: what the various "world leaders" are saying and pretending to be doing about it. Here, a simple observation will suffice: have they done anything about it so far? Since 1992, when the Kyoto Protocol was negotiated, have global industrial greenhouse gas emissions gone up or down? The answer is that they have gone up and by a stunning amount! Therefore, while climate change can be a legitimate focus, unless you can offer a compelling argument for why "this time, it will be different," climate change politics is purely a distraction.

But what, you might ask, if you actually like your distractions and aren't at all fond of the idea of giving them up? Of course, all things are good in moderation, and this includes being focused on eliminating distractions. You should probably work on eliminating those that are not at all useful or pleasant and the ones that hinder you in achieving your goals while keeping the ones you truly enjoy, be they the study of ancient civilizations, flower arranging or chess. And perhaps you should allow yourself to be distracted by utter nonsense—once in a while, to avoid taking yourself too seriously.

Problems of scale

THE PROBLEM OF scale is largely the result of a habit of thought: immersion in the technosphere teaches us to think that scale is arbitrary. We are taught that we can measure things from the nanometer to the light-year, that we can see continents move or travel at several times the speed of sound, and that one person can be a billion times wealthier than another and still be just a normal person rather than a person whose buttocks encompass the globe.

None of these things is actually true. We are not able to directly see with our eyes anything much smaller than a tenth of a millimeter. Only a few of us can run a four minute mile. And without powerful social machines providing support for concentrating unlimited amounts of abstract, notional wealth, none of us could be that much richer than the next person because that next person would come up to us and say, just as often happens in the normal course of human relations, "You clearly have more than you need; let me help you get rid of the excess."

But since we are taught that scale is arbitrary, we are led to believe that bigger is always better—richer, more powerful, more impressive—and constantly overshoot the optimum size of just about everything. The place where this causes the most harm is in the scale of human society. Clearly, the optimum is larger than one: as individuals we are weak, virtually defenseless; as a group we can be almost arbitrarily strong. But our individual power increases as the size of our society increases only up to a point, and the optimal size is surprisingly small—somewhere between 100 and 200 individuals, with 150 often regarded as best. This was discovered by the sociologist Robin Dunbar and is known as the Dunbar Number. Beyond this point, the larger the group is allowed to grow, the weaker each of the individuals within it becomes, on average.[6]

So, what does this mean for us, living in cities with populations in the millions, in countries with hundreds of millions, on

---

6   The Dunbar Number relates to cohesive groups that trust each other and are able to act as a unit; in a stable, benign social environment loose networks of individuals based on limited trust can be quite a bit larger without exceeding optimum size.

a planet with billions of us? What it means is that those billions cannot be signified by the word "we" in any useful sense of this first person plural pronoun—they are just numbers, at a scale that cannot be directly comprehended using our human senses. Platitudes such as "We are the world" and "Peace on earth" may have a pleasant ring to them, but to anyone with the slightest inkling of what actually goes on in the world they are a bit hard to stomach. Those billions don't exist in any tangible sense; they are but shadows projected onto the walls of the cave to which we are chained.

And who does exist? Those who aren't strangers to us: those we know personally and have personal relationships with. They are family in the most expansive sense of this word—taken to include everyone who can be included (not necessarily just blood relations) without that word starting to lose its meaning. It is all the people who can never in good conscience exclude you without an absolutely outstanding, ironclad reason; who will grant you favors in spite of not owing you any; who will treat you better than they will treat any stranger whether you deserve it or not. In an extreme situation, these are the people who will willingly give up their lives to defend yours and expect the same in return.

The existence of such families, or clans or tribes, is a direct affront to the technosphere, which prefers us to be defenseless, atomized individuals, helplessly dependent on impersonal social machines and technological life support systems. The battle lines become drawn the moment you join with others in an effort to form such a community. This becomes obvious rather quickly, because when you start working directly with other people in circumvention of any official arrangements, you will soon find out that you are breaking the law. Trade a bunch of radishes for a bunch of parsnips with a neighbor, and if you fail to report this barter transaction to the US government, you become guilty of income tax evasion. In fact, a good litmus test for finding out whether someone is family or a stranger is whether that person is willing to break a few laws for your sake. If not, then you are not working with a true friend but with an agent of the technosphere who has infiltrated your midst. Now, lots of groups take this too far and require a new recruit to commit a serious crime as part of a rite of passage for gaining acceptance, but the point still stands: if doing the right thing is illegal, then requiring some amount of illegal behavior offers a way of avoiding split loyalties.

The best way to solve the problem of scale is to solve it on a personal level, by finding your 150[7] or so closest people and by excluding the rest from your life to the greatest extent possible. This is a very large behavioral shift for most people. We are taught not to work with family and friends because that is considered nepotism and favoritism; yet that's what usually works best. We are taught to trust impersonal institutions such as banks, insurance companies and investment companies, which are all riddled with corruption and fraud and are far too powerful for even the government to stand up to, while at the same time we are taught to mistrust the people who are the closest to us and who would work the hardest to earn our trust if offered the chance. We are told to do business with complete strangers, because they supposedly care about their public reputation or because the police and the courts will come to our defense if they cause us harm or defraud us. All of these bad habits can be hard to break all at once, so it is best to do this gradually—one personal relationship at a time.

Once you, and the few people who really matter to you, solve the problem of scale at the personal level, the problem of scale at the level of the huge—city, country, planet—remains, but it is no longer as much of a concern for you, because you no longer depend as much on things outside of the very local scope of family and friends. And if you do find that you need to concern yourself with these external entities, you can confront them not as a weak,

---

7   This subject is explored in much more detail in *150-Strong: A Pathway to a Different Future* by Rob O'Grady, Club Orlov Press, 2016, https://www.amazon.com/dp/1523676523]

defenseless individual, but as a strong, cohesive group. If your town or your city fails any one of you, then the local officials suddenly find that they have a much bigger political problem on their hands than they bargained for. If your country fails you, then your tribe can stage an exodus to a more promising one. You will always know what's really important to you, and none of these out-of-scale entities will matter more to you than your 150.

Lifehacks

AND NOW WE come to the crux of the problem: beyond picking away at the technosphere using the technique of harm/benefit analysis, how can we wage a pitched, victorious battle against it? In order to answer this question, we need to look at the strategies that the technosphere uses to control and enslave us, and then learn some counterstrategies that have been shown to be particularly effective.

The overall strategy of the technosphere in ensnaring and enslaving individuals boils down to just one tactic: making you pay. In order to simply live—to inhabit the planet, as is your birthright as an Earthling—you are forced to pay rent or a mortgage. To eat food, some of which grows wild, while the rest can be cultivated with a modicum of effort, you have to pay for it. To survive in a polluted environment, in crowded and stressful conditions that are conducive to spreading disease, you need access to medicine—and, of course, you have to pay for it, whether directly, by buying insurance or by paying taxes.

And in order to afford all of these expenses, you have to earn money—and that's where you relinquish virtually all of your autonomy, because holding on to a job requires you to do as you are told, or obey written instructions that have been handed to you. Most of the jobs that are available are superfluous—they could be done either by robots or by near-slave labor in some impoverished country, or not done at all—and they exist in order to exercise control over you, to deprive you of free time and of the ability to do anything that is not in the interests of the technosphere.

The biggest expense is usually the rent or the mortgage. Everybody needs a place to live—nobody wants to be homeless—and so this is a major opportunity for the technosphere to exact its tribute. The "home"—a word which, in the context of financial enslavement, takes on a sinister tone—should really be called "housing." Housing is strictly regulated to be a certain kind of permanent, stationary box and is required to be connected by certain required umbilicals to public utilities in order to increase your level of dependency. It is designed to be sufficiently expensive to make it unaffordable for a significant number of people and to make homelessness enough of a social problem to scare people into submitting to all manner of indignities in order to avoid it.

The next net to ensnare you is composed of consumer-level products. Here, the technosphere's strategy is to take something that can make you independent—a tool that you pay for once and use for the rest of your life, possibly bequeathing it to your children—and make you replace it as often as possible by building in planned obsolescence—ideally after each use—by replacing it with a single-use, disposable item. This does not happen by accident, but is a technique taught in business schools and industrial design courses. Here, your task is to avoid consumer-grade products, instead relying on industrial-grade products (because industrial customers have far more clout in deciding what time frame is appropriate for depreciating a piece of equipment). You can also improvise your own designs with a view to making them last a long time and be maintainable.

Next are commercial services. Here, the idea is to make you depend on paying people for as many services as possible. Jobs are designed to keep you busy, so that you don't have time to shop at farmers' markets or specialty shops that offer local produce, or to grow your own food, but instead have to drive to the supermarket. To save even more time, you skip cooking and buy industrially prefabricated meals laced with chemicals and full of empty calories. Instead of helping yourself and those around you in as many ways as you can, you are encouraged to specialize and only do the one job for which you are trained.

But it is far more efficient from a personal point of view (and supremely inefficient for the technosphere) when an entire neighborhood decides to de-specialize and to provide services to each other in ways that do not involve money. These include, in no particular order, lawn care and landscaping, cooking and catering, construction and renovation, farming, fishing, hunting and gathering wild foods, equipment installation and repair, automotive repair, home aid, child care, tutoring, haircuts, garment making, interior decoration, counseling, burial services, first aid, prenatal counseling and midwifery, transportation services, hospitality services and entertainment. A community that internalizes and de-commercializes all of these services takes a gigantic bite out of the flank of the technosphere. It also gains a great deal of peace of mind, health, free time, community spirit and happiness.

And then there a few tricks—or lifehacks—which are so powerful that they are worth mentioning separately. They provide methods of escape that are within many people's reach.

Boats

One of the most powerful and pleasurable methods of avoiding paying rent or mortgage is to give up on land-based dwellings and move aboard a boat. Although the boating industry is geared toward sport and recreation, catering to well-to-do people and well-funded sportsmen with very expensive, shiny, disposable plastic toys, there is also an entire community of people who live aboard and travel aboard older, quite affordable boats, which were built before the boatbuilding industry discovered ways to build in planned obsolescence. The live-aboard communities that spontaneously come together around marinas, mooring fields and anchorages tend to be very supportive of each other, trading advice, helping each other with repairs and arranging swap meets to save on parts.

The cost comparison between landlubber and live-aboard lifestyles is stunning. I have spent quite a lot of time in Boston, which is one of the most expensive cities in the world, and there the break-down is as follows. A one-bedroom apartment will run you around a quarter of a million dollars; a perfectly habitable boat can be had for less than $20,000. Rent for a one-bedroom in a neighborhood with foul air, in an ugly building infested with bedbugs, featuring a view of a parking lot, will set you back around $2,000 a month; a slip at a marina with a pleasant sea breeze, a "million-dollar view" of the harbor and the city skyline, a gym and a swimming pool will set you back $700 a month. Landlubbers pay a lot for vacations: airline tickets, hotels, rental cars, restaurant meals and so on. Live-aboard folks can just sail to a secluded anchorage and fish, swim, relax, eat home-cooked meals and generally enjoy themselves.

Landlubbers tend to accumulate junk—because they have a place to put it. They buy appliances and tools they hardly ever use, clothes they hardly ever wear, furniture and collectibles. People who move aboard put all that stuff in storage, and then, a few years later, realize that they don't miss any of it and get rid of it. On a boat space is limited, and people are forced to be very selective about which possessions they wish to hold on to. If you find that in order to buy something you have to decide what to get rid of, you auto-matically tend to buy fewer things. Instead of accumulating junk, you suddenly find yourself accumulating money.

Having so many fewer things on which to spend money makes it possible to work much less and to sail a lot more. And this buys you plenty of free time to relax, to exercise and to socialize with other live-aboards. Mood and health improve and, with a bit of effort, once impossible dreams start coming true. And if things don't go as well as you'd wish, the boat is there to help. If you lose your job and can't find another, and can no longer continue paying rent or mortgage on an apartment or a house, you end up homeless rather quickly. On a boat, you can move to a cheaper marina, or you can move out to a mooring field (your own mooring will set you back $300 or so) or, if you are really broke, you can anchor and get to shore by dinghy. In Boston you can anchor free of charge forever in full view of a row of million-dollar condos.

And if life in the big city gets to be too much (it is, after all, full of artificially busy, stressed-out people who have no time for each other—or for you), then you can set sail and eventually get to a place where everybody is always relaxed and smiling, where a cold beer at a bar on the beach costs $1, and where on most days you can find a grilled barracuda to have for dinner for not much more than that, without even leaving the beach.

Tiny houses

Not everybody can move aboard a sailboat and eventually sail away to a tropical paradise. Some people are stuck in a landlocked place that does not contain any large bodies of water or are too far north to live on a boat year-round. Others can never get over sea-sickness, or become anxious when they lose sight of land, or simply don't like boats. For these people, similar advantages can be realized by constructing and moving into a tiny house. Tiny houses are often built on a trailer, so that if local conditions change for the worse they can be towed to a more welcoming locale. If you need to be near a city, you can rent a spot for it; if you want to live close to nature, you can tow it to the middle of a meadow, a lakeshore or the edge of a forest.

There is a very healthy tiny house movement already and plenty of literature available on this subject for anyone who is interested. To briefly summarize their benefits: tiny houses offer the second most effective way, after boats, to avoid having to pay an exorbitant amount of money for a place to live; they are relatively cheap to build and maintain; they can be built by any reasonably handy person in a short period of time; their design can suit the individual fancy of the builder; often, they can be built from recycled or salvaged materials; and they are relatively easy to move from place to place.

It bears pointing out that there is a middle ground between tiny houses and boats: the houseboat. This can be nothing more than a tiny house on a barge. Tiny houses are still considered houses, and local authorities may tax them as such or claim that they violate some zoning restriction or a local ordinance and fine you. The solution, of course, is to simply move it to where the grass is greener—or where you haven't yet overstayed your welcome. But with houseboats there is a foolproof loophole to exploit: if some authority claims that it's a house, you put an outboard motor on it, motor around for a bit and tie up at the dock again. Voilà, it's been demonstrated to be a motor vehicle, not real estate. And if local authorities claim that it's a boat and that living aboard boats isn't allowed, you take the engine off, and voilà, it can't move under its own power and is therefore a house, and you can go on living on it because nowhere does it say that it is illegal for a private residence to be able to float.

Yet another variation on the theme is a houseboat that sits on dry land. This may seem strange, but houses do have basements, and at a minimal additional expense basements can be poured as above-ground concrete barges instead of as buried five-sided boxes. This has several advantages. First, such a house is technically a boat (stored on dry land) and is regulated far more leniently than a house, making it far easier to set it up for off-grid living. Second, it can be located in a flood zone: when the water rises, the house floats up on pilings; when the water recedes, it settles down again. Third, with such a house, a flood can be seen as an opportunity rather than a danger, giving you a chance to move it to a nicer location.

Free data

We happen to be living through an age of digital information, which probably won't last forever since it relies on short-lived devices that use up prodigious amounts of energy and lots of scarce nonrenewable resources such as lithium, tantalum, gallium and many others. But while it lasts, we might as well take advantage of it. Digital information has a particularly useful property: it can be copied any number of times without any quality degradation at all, because a 1 remains a 1 and a 0 remains a 0 no matter how many times you copy them.

But what the technosphere wants to do is to charge you for access to information, via a wide variety of schemes based on intellectual property laws. But there is nothing particularly intellectual about a zero or a one or any number of them in any combination.[8] Intellectual property laws are perversions of laws that were created long ago to protect the rights of intellectuals—authors, artists, inventors—hence the name, but they have since been subverted to serve the needs of the technosphere's stakeholders and legal representatives.

However, these laws are hilariously easy to circumvent and quite difficult to enforce. And so it turns out that whatever it is you want you can have because it is available for free—as long as it is in digital form. Some people go into high dudgeon upon hearing such things, holding forth that this is reprehensible, illegal, criminal, immoral or any and all of the above, and further claiming that if everybody did this it would destroy the global economy. But this is a point that cannot be proven; for all we know, it could just as easily improve the global economy by reducing inequitable concentrations of capital. You should certainly do your best not to trample on the rights of intellectuals: individual authors, artists and inventors should be able to reap some sort of financial reward from their efforts. But when it comes to the "intellectual" property of manifestly nonintellectual corporate entities, the moral argument holds no water at all.

Free data conveys numerous advantages. Entire libraries of books, music, video and art can be stored on a stack of flash drives and made available to an entire community via the local wireless network. Software libraries, made available in a similar manner, can support a wide range of activities, from typesetting books to 3D-modeling and design. The same community-wide wireless

---

8   In fact, it can be argued that a digital representation of every work ever to be created can be found somewhere among the infinite number of digits of the irrational number π, meaning that computing π to arbitrary precision automatically violates intellectual property laws.

network can provide local voice-over-IP mobile phone service. Community networks can be interconnected using line-of-sight radio transceivers, which now cost around $250 and provide 100 MB/s speeds with better than a 100-kilometer range. A similar link can provide the entire community or cluster of communities with internet access. All of it can be powered using solar panels. Such technology allows constellations of scattered communities to have access to a mass of knowledge and information resources, and to remain in contact with each other and the outside world, without having to pay rent once the equipment has been paid for. In terms of harm/benefit analysis, there is some harm (the equipment, in being manufactured, did use up nonrenewable natural resources and did cause pollution), but the benefits are quite remarkable.

Free-range children

Strong, self-reliant families are antithetical to the technosphere's quest to supply itself with a steady stream of atomized, alienated individuals, easy to control and subservient to its needs. The strategy of weakening and destroying families is pursued using numerous tactics. One particularly egregious example is the tactic of depriving families of the ability to earn a living wage, then providing them with public financial support in a manner that makes fathers superfluous by granting greater levels of support to single mothers.

But an even more important tactic is applied almost across the board in a two-step program. Step one is to make it impossible for parents to bring up and teach their own children, by filling up all of the parents' time with busywork on the one hand and by establishing artificially inflated educational standards that the parents cannot match on their own on the other. Step two is to rob children of their childhoods, forcing them to attend pre-school, then school for an interminable number of years, during which they learn very little that is of practical value and are instead prepped for even more schooling—in college.

To make sure that as many as possible end up in college, both the parents and the would-be college students are subjected to a barrage of propaganda telling them that without a college education they would not be able to achieve "success." This narrative runs contrary to reality, which is that barely half of college graduates are now able to find jobs that have anything at all to do with their degree. Yes, most of the degree-holders in computer science and engineering can still find jobs (which many of them eventually grow to hate), but these are the specializations that are particularly important to the technosphere. Most of the rest of the students are simply being taken for a ride. And the purpose of that ride is not to give them an education but to saddle them with staggering levels of student debt, turning them into lifelong debt serfs.

But if you have been keeping up with the program of wresting control from the technosphere, the parents among you should be able to gain plenty of free time to spend with your children, to bring them up yourselves and to teach them a wide variety of skills which they can practice directly to help themselves and those around them. Children love to play and should be allowed all the playtime they want, but they also respond remarkably well to being given a chance to contribute.

There is no reason why five-year-olds can't be taught to water the garden and pick vegetables for dinner, why six-year-olds can't knead dough for baking bread, or why seven-year-olds can't be taught to fish on par with adults. Those with a mechanical aptitude can be taught to repair and overhaul various pieces of equipment from age nine or ten.

Then there are all those things that children sometimes insist on learning—a foreign language, chess, history—and this is the best and most successful type of learning because the motivation for it comes from the inside. Some children—a minority—have an innate curiosity and start experimenting on their own; these are the only ones that should be taught science and medicine. Other children—again, a minority—discover literature early on and cannot be kept from reading, keeping a journal, composing poems and so on; these are the only ones that should be taught literary skills. The same is true of painting, music, taxidermy and sculpture—all the things that are worthless if the lack of significant talent causes a student to be mediocre at any one of them. It is very important that schools, which are full of misguided educators who believe that everybody should be taught everything, aren't allowed to get in the way of any of this valuable learning.

Then comes puberty, followed by a lengthy time-out during which an entirely different sort of learning takes place—though just as important—and school should not be allowed to get in the way of it either, although having productive activities to engage in as a distraction from raging hormones is certainly helpful. Teenagers are often perfectly happy assisting with carpentry, cooking, painting, sewing, construction, gardening, plumbing, animal care, electrical work and much more. The positive end result is young adults who have had happy childhoods of play, productive contribution and self-motivated learning, and who emerge into the world in possession of a multitude of practical skills. At an age when their peers are recuperating after many years of daily confinement in the classroom, eager or not so eager, to start applying for jobs, but really only knowing how to push buttons and scribble on bits of paper, they can apprentice themselves to a master and at the conclusion of the apprenticeship start working for themselves.

Let's make one thing perfectly clear: official schooling is not necessarily about learning anything useful, but it is definitely all about learning obedience. The product of the educational process, with its largely arbitrary content, standards and rules, is a model prisoner of war: disciplined, obedient, indifferent, numb. Both the more vulnerable and the more independent-minded children tend to have their spirits crushed. Their self-esteem becomes dependent on their ability to compete in games not of their choosing and on the approbation they receive from people they do not love. Girls, being of a more docile nature, do better at surviving school than boys who, being more rebellious, often end up damaged for life, emerging not as men but as badly behaved girls.

I found myself in that place more than once, and what saved me was the simple understanding that the grade I get is the grade I give—to the instructor. I was generous and gave out far more A's than my teachers and professors deserved; I doubt that I would be so generous with them now. As a university graduate with a couple of advanced degrees, I have this to say: "Dear teachers and professors, thank you for all the 'time served,' but I regret to inform you that most of you have flunked my education." To be fair, a few were excellent and a fantastic help, but in two decades of classroom/lecture hall confinement I can count them on the fingers of one hand and consider them to be the exceptions that prove the rule that formal, official education is largely a waste of time.

If you feel that wresting control is too arduous a task for you, then think of your children (if you have any). It seems eminently unfair to make them follow a failed plan and expend the precious years of their childhood and youth in a futile effort to become cogs in a wheel that is going to either grind to a stop or fly apart. If you don't want to do this for yourself, then do it for your children!

9
=

THE GREAT TRANSITION

WOULDN'T IT BE nice to be able to say that everything you have
read in this book so far is entirely optional, purely for your entertainment? Then you could finish this book and go on with your life
as if nothing happened, perhaps taking some of what you read into
account, perhaps not. Unfortunately, this is simply not the case.

Regardless of whether this prospect pleases you or distresses
you, the technosphere is going to fail you. There are simply not
enough easy-to-exploit, concentrated, conveniently located nonrenewable natural resources left to sustain a global industrial order.
As has been laid out in the preceding chapters, the technosphere,
as a single, integrated, emergent intelligence, is in extremis. As
it enters its death agony, its previous depredations may come to
seem mild compared to what happens next.

A dire but realistic scenario is likely to include some or all of
the following:

- We should expect widespread nuclear meltdowns, which we will
  be unable to prevent because of societal disruption and the disappearance of the vast industrial resource base needed to keep
  nuclear installations safely under control. This danger is currently
  quite elevated in the Ukraine, where, due to a collapsing economy and chronic shortages of natural gas and coal, nuclear power

plants are being used to supply power for peak, intermittent loads rather than for base, constant loads—a function for which they were not designed. It was just this sort of dangerous rapid cycling of a nuclear reactor that caused the meltdown in Chernobyl, in the Ukraine, four decades ago.

- We should expect even more profound climatic disruption and toxification of the biosphere to result from increasingly desperate efforts to exploit ever more polluting, ever less efficient sources of fossil fuel energy. Already the almost complete lack of new conventional oil and gas discoveries has forced the global energy industry into exploiting marginal, environmentally devastating and financially unprofitable resources such as shale oil and gas, tar sands and extra-heavy crudes such as the ones found in the Orinoco Belt in Venezuela. But even these desperate measures have been predicted to fail within the next decade as energy returned on energy invested (EROEI) falls below the 10-to-1 threshold—the level needed to maintain large-scale industrial capacity for producing and consuming fossil fuels.[1]

- The sweeping away of democracy and human rights in a desperate effort to maintain control over increasingly dissatisfied, distressed and rebellious populations. Various countries around the world are experiencing a continuing population spike even as resources dwindle. In what is still being optimistically referred to as the developing world, large urban populations have grown up on food derived from natural gas via nitrogen fertilizer. The process of increasing agricultural yields by spiking the soil with chemicals has been misnamed "the green revolution"; as resource depletion runs its course, yields will plummet, food prices will rise, and the populations this revolution has created will face hunger and starvation. The sprawling slums they inhabit will become ungovernable, resulting in high levels of violence and unleashing floods of

---

1   For a comprehensive summary of the situation, see
http://www.thehillsgroup.org/.

migrants who will inundate and destabilize other countries. Rapid climate change will make this problem significantly worse. Already we are seeing the effects of climate change-related droughts in Syria; there, civil unrest (exacerbated by foreign intervention) has led to a five-year-long civil war, widespread devastation and millions of migrants who have now flooded into western Europe accompanied by tens of thousands of radicalized Islamic terrorists, giving rise to a wave of terrorist attacks.[2] Liberal democracies are ill-suited to countering such threats and will either be forced to resort to more authoritarian methods of governance or will turn into failed states.

- Even the still relatively stable and prosperous nations run a high risk of political, financial and economic instability as previously successful economic policies stop working with nothing to replace them. Everywhere in the world economic growth is stalling out—a fatal development for a debt-based financial and economic system predicated on endless growth. Wherever increasing levels of debt have been used to drive economic growth, continuous growth has become necessary to make debt repayment possible. Around a decade ago growth mostly stopped, making debt repayment impossible, but to prevent a wave of bankruptcies and defaults Western and Japanese central banks made the fateful decision to roll the debt over perpetually and to allow it to expand continuously. As a result, the financial system has gradually been transformed into a pure pyramid scheme that will collapse as soon as confidence wavers. This process has been cannibalizing the real economy of jobs and services to prop up tottering financial pyramids, exacerbating wealth inequality, punishing both younger generations and retirees and destroying the middle class. As it runs its course, public discontent is guaranteed to rise to extreme levels.

---

2   It doesn't help matters that these terrorists are often put to use by the very same Western government agencies that are charged with fighting terrorism, in an effort to justify their existence and expand their budgets and roles.

In the US—which is the epicenter of the global financial explosion of unrepayable debt—there are now militant groups with hundreds of thousands of heavily armed members (many of them battle-hardened military men returning from tours of duty in Iraq, Afghanistan and elsewhere) who openly deny the legitimacy of the US government and aim for its violent overthrow. In the summer of 2016 some of them launched an assassination campaign against the police. If this particular wave of revolutionary terror follows the patterns of previous ones, the campaign of home-grown terror will gradually begin to target not just policemen but also politicians, members of the judiciary, military leaders and other public figures essential to the functioning of government bureaucracy. In this context, those who plan to remain reliant on government services could be in for a nasty surprise.

It would be comforting to think that most of those reading this book will be able to somehow escape all of this instability and unrest. Of course, some will remain unscathed. Pockets of stability and continuing economic activity may persist in a few countries and regions—the ones that have the unusually lucky combination of a relatively stable climate, an abundant natural resource base, a self-sufficient local industry and no population pressures. They may be able to avoid instability and social collapse. Some of them may even succeed in effecting a peaceful, orderly and gradual transition to a post-industrial age. It certainly makes sense to try to identify such countries and regions ahead of time and to attempt to find refuge there. But only some of us will be able to pursue this strategy.

As for the rest of us, a number of choices still remain:

I.   Use what little time remains to shrink the role the technosphere plays within the life of your family and your community. Reclaim as much autonomy, self-sufficiency and freedom as possible. Attempt to insulate yourself to the greatest extent possible from the ravages of the technosphere as it enters failure mode. Spare yourself and your children the lengthy process of earning credentials to qualify for jobs which will no longer exist and invest instead in acquiring useful skills that can make you self-sufficient in building and maintaining shelter, growing food, taking care of your medical needs, providing for your defense and security, and much more.

2. Psychologically prepare, train and equip yourself to be able to roll with the punches and perhaps even to take advantage of new situations by learning to fish in murky waters. After all, bad times usually offer big—though unethical—opportunities to make money. Right now huge fortunes are being made selling arms to terrorists in Syria, transporting migrants across the Mediterranean in inflatable boats, producing and marketing Afghan heroin under the watchful eye of NATO troops and Western bankers... In the US, gun manufacturers and dealers are making healthy profits even as mass shootings have become an almost daily occurrence. Such opportunities are sure to be plentiful for those unscrupulous enough to exploit them. But keep in mind that even if you manage to amass a great fortune and find ways to yank it out of financial institutions and instruments before they become defunct, your hoard may turn out to be so difficult to defend that it will become more of a liability than an asset and of quite limited value, because the things you will want to spend it on will no longer be available once industry largely shuts down.

3. Give up on preparing and simply try to live each day to its fullest, as if it could be your last, exercising as much freedom as the situation allows you at each step. After all, the more uncertain you are about the future, the more insistent you should be that you need to lead a meaningful, satisfying, memorable life in the present. If you are assiduous and steadfast in pursuing this plan, yours is sure to be an interesting voyage of discovery, and it may by pure chance take you to a more interesting and promising place than any you might find through careful research, planning and preparation.

Depending on your situation, your character and your abilities, some combination of these choices will apply to you. There is no ideal approach and no path to perfection. Just surviving is hard enough, and what survival demands more than anything else is flexibility. Each of these choices has its merits and demerits.

There is one more choice for you to consider—and dismiss as promptly as you possibly can. It is to continue to do as you are so often told to do: work or study or simply waste your time along with everyone else's, sitting on an office chair in a temperature-controlled building, commuting by car, shopping at supermarkets and department stores supplied by diesel trucks and container ships, paying for it all using plastic cards tied to a global financial network, maintaining a high level of debt, paying into and expecting that you will someday get to collect retirement... and believing that your children will be able to find jobs when they graduate and repeat this same cycle, endlessly. This choice has one obvious merit: it doesn't require you to change anything. It doesn't even require you to think! But given how far along you are in reading this book, if any of it has made a real impression, then you will probably find passively waiting for the inevitable much too nerve-wracking.

Let me propose a better approach—one that can offer you some peace of mind. Let us separate the challenges we face into **long-term risks**, which in the short term we can do nothing about and **short-term risks**, which we can usefully start preparing to face right now. Then we can immediately start going about making the sorts of preparations that make the most sense as an immediate investment in our safety and peace of mind. Such activity, you may be surprised to find out, is not in any sense radical or extremist but is specifically recommended by official organizations such as FEMA[3] in the US and the emergency ministry in every country that has one, down to detailed lists of items you should purchase and stockpile. According to these agencies, it is your responsibility to be ready to evacuate at any moment, bags packed with all of the immediate necessities for yourself and your family, including

---

3   Federal Emergency Management Agency

medications, cash, documents, warm clothes, etc. In addition, they recommend that you maintain in your home several weeks' or months' worth of food, water and other necessities. It is the job of these agencies to help you in case of emergency, and it is most efficient for them to teach you how to self rescue instead of having you wait around to be saved. It is an awful lot more effective for people to prepare for their own disaster relief by thinking ahead a little.

As far as the long-term risks, here we have to recognize what kind of world we are going to bequeath to our children and to teach them what to expect and what will be expected of them. While it would be a great disservice to our children if we fail to explain to them what sort of future they face, it would be equally inadequate of us to demoralize them by offering explanations that are purely negative and paint a bleak and hopeless picture. A much better approach is to teach them what they will need to know to be safe, healthy and happy in spite of all of the problems they are likely to encounter and to give them opportunities to test their knowledge and practice their skills.

The point is, you must *show* them how to live and not just teach them theory while contradicting yourself in practice, because cynicism, hypocrisy and insincerity are adult character traits that children have no way of appreciating. Children learn by imitating our behavior, and if it contradicts our thinking then at best they learn to simply ignore what we say and at worst become troubled by it. Suppose you teach them about the environmental devastation they will witness during their lives, and explain to them that it is being caused by burning fossil fuels, and that during their lives fossil fuels will disappear altogether with nothing to replace them... while continuing to burn hundreds of gallons of heating oil to heat an oversized house, driving all over creation in an oversized vehicle, jetting off to the tropics on brief winter holidays and going on shopping sprees to buy on a whim things you don't need. Then what you would be teaching them is that you can't be trusted.

And this doesn't help them; instead, it damages their spirit. It is better to have an ignorant fool for a parent than a well-informed hypocrite because being a fool is not a moral failing. Fools deserve pity and mercy; hypocrites—neither.

Even if you are a model parent who "walks the talk" and provides future generations with both instruction and hands-on training on how to deal with all of the foreseeable risks, a problem remains: the surrounding society will remain mired in denial, will do nothing to prepare and will sit around waiting for the return of growth and prosperity, any moment now, as endlessly promised by their elected and unelected leaders. How are the children, who, like most children, seek acceptance and approval through mimicry and conformism, going to be able to cultivate a particular mindset and pursue a separate agenda as part of the family while continuing to fit into a larger group that is likely to be unreceptive, and even hostile, to any such ideas? The answer is that your kids need to be taught method acting. Within the inner circle of family and trusted friends they can be themselves, while among strangers they should be taught to display an artificial persona specifically designed to achieve various explicit objectives. Having been raised in the USSR in a family of secret dissidents, I speak from experience. I was taught from an early age the value of privately shared knowledge and that truth is a powerful secret that has to be prevented from leaking out into an outside world ruled by ignorance and lies. It is a hard lesson, but learning how to act makes it easier.

Long-term risks

NO MATTER WHAT we do, we and our descendants are sure to face a formidable array of problems, because we will be living in a world that has been ravaged by the technosphere, disrupted and transformed in ways that will affect us for the remainder of our existence as a species. There is even a chance that these effects will result in our premature extinction. We know that all of the things listed below are in fact happening; what we don't know is exactly how quickly, or just how bad the situation will eventually become. The list below is the stuff of depressing dystopian novels. Much of it beggars the imagination; not only has none of us ever experienced such conditions during our lives, but they lie outside of what us sentient hominids have experienced during any of our millions of years of existence, leaving us unprepared on every level—economic, political, cultural and biological.

- Ocean levels rising by as much as 150 meters (500 feet), putting the coastal cities, in which two-thirds of the population lives, either partially or fully underwater, while storm surges from increasingly powerful typhoons and hurricanes, and meltwater pulses from rapidly melting glaciers render investments in coastal flood defenses meaningless. The latest forecast, which is probably still too conservative, is for a 3-meter (9-foot) rise in ocean levels by 2050–2060, which will already be sufficient to permanently flood a large number of coastal communities.[4]
- Average temperatures ultimately rising by as much as 15°C from their pre-industrial levels, resulting in heat waves that take down electric grids, along with the air conditioners they power, depopulating major metropolitan areas through death from heatstroke, while forests and peat bogs dry out and burn up, prairie turns to sand dunes and previously agricultural regions turn to desert
- The disappearance of winter snow pack and mountain glaciers which feed river systems on which major agricultural areas depend for irrigation, resulting in depopulation due to mass starvation in many countries around the world
- Abandoned industrial installations of all kinds leaking radioactive contaminants and chemical toxins into the environment, making

---

4   These numbers were quoted by Margaret Davidson, NOAA's senior advisor for coastal inundation and resilience science before an audience of risk management professionals at RIMS 2016 conference in San Diego, California.

it too dangerous to live near them or downwind or downstream from them

- Increasing disease loads and epidemics of previously treatable infectious diseases, as bacterial evolution outstrips the efforts to develop new antibiotics, while the warming climate causes tropical diseases to spread to the formerly temperate zones

- The oceans becoming too radioactive to fish as dozens of nuclear installations built along sea coasts become inundated with salt water from the rising seas. Given the huge volume of the oceans, the radionuclides will still be quite diffuse, but they are concentrated by living organisms, reaching high concentrations in the predators, highest in the top predators—humans.

- The oceans absorbing carbon dioxide from the atmosphere, produced by burning fossil fuels, will become so acidic that they will start dissolving coastlines composed of limestone, eating away at them and turning them into impassable landscapes riddled with crevasses and sinkholes

- Most natural resources, such as metal ores and minerals, good soil, forests, fisheries and populations of wild animals becoming too depleted to be of use to people attempting to exploit them using nonindustrial methods

- Populations in distress everywhere, most of them entirely unaccustomed to surviving within the natural environment with the help of locally available materials, simple tools and their own wits

This list may be enough to make certain people curl up into a fetal position and rock back and forth weeping. But I hope you will agree that this is not the right approach. After all, we aren't dead yet, so we might as well keep trying to stay alive. There is probably no longer anything we can do to prevent any of these problems. But we can take steps to protect ourselves from their effects.

For example, we can learn how to detect unsafe levels of chemical or radioactive contamination—on our own, without government help. This turns out to be quite necessary, because whenever there is a serious nuclear or chemical accident all governments everywhere seem to do one and the same thing: they lie. The Soviet government lied about the extent of radioactive contamination after the Chernobyl disaster in the Ukraine. The Japanese government lied, and continues to lie, about the ongoing radioactive contamination caused by the multiple meltdowns at Fukushima Daiichi.[5] Other governments are now lying about the safety of radioactive imported Japanese seafood and rice, and instead of banning it introducing regulations to prevent its source from appearing on the label. Michigan state government officials have lied relentlessly about the level of lead contamination in drinking water in Flint, Michigan. Government officials at every level have lied, and continue to lie, about the harmful effects of the dust from the demolition of the three skyscrapers at New York's World Trade Center on 9/11 and about the toxic and radioactive contamination being leaked into the environment by the oil and gas industry. Thus, whenever there is a serious accident involving chemical or radioactive contamination, there is every reason to expect that the government is going to lie about it, and so you have to learn how to find out the truth yourself by making your own observations and measurements.[6]

Preparing to cope with long-term risks is an intergenerational project spanning many decades. Some of the worst ravages of a destabilized climate may not directly affect you, but they will

---

5   Most politicians instinctively tend to favor appearances over substance. After each of these two nuclear disasters, Soviet and Japanese governments both insisted that the reactors were being cooled, ignoring the evidence that the reactors no longer even existed—they melted down, their nuclear fuel disappearing into the ground below.

6   This topic is covered in detail in NRBC: *Surviving Nuclear, Radiological, Bacteriological and Chemical Accidents* by Cris Millenium and Piero San Giorgio, forthcoming in 2017 from Club Orlov Press.

certainly affect your children and grandchildren. While being aware of these long-term risks is important for forming a realistic set of expectations about the future, it is difficult to take them into account in the way you live your life day to day, or to directly translate this awareness into decisive, purposeful action. It is also very difficult to form a consensus, even within a single family, over possible future impacts whose extent and timing remain uncertain. But there is another set of risks that are indisputably real and that exist in the here and now. Numerous kinds of short-term crises and disruptions, at every scale—personal, social, national, global— may have a low probability of happening at any given time, but your exposure to these risks is constant and cumulative over time. Therefore it makes sense for you to deal with them here and now.

Short-term risks

IF YOU DO something that isn't safe, but only do it a few times, then there is a good chance that you will get away with it. But if you make a habit of it, then chances are you won't. For example, if once or twice you take the risk of getting behind the wheel after you've had a few, the probability that you will run over and kill somebody is quite small. But if you take that risk every single weekend, that probability gradually turns into a certainty.

If the probability of a certain behavior resulting in a bad outcome is some very small number p, much closer to 0 than to 1 (where 0 is an impossibility and 1 is dead certainty) and n is the number of times you engage in it, then the cumulative probability of that bad outcome can be calculated as follows:

$$\sum_n p = p_1 + p_2 + \ldots + p_n$$

For some number n the probability of a bad outcome becomes greater than 1.

The distinction between constant risk exposure and

intermittent or incidental risk exposure is a vital one. It only makes sense to prepare to cope with intermittent or incidental risks if the probability of a bad outcome is quite high. But with constant risk exposure it makes sense to prepare even if the probability is quite low. For example, you might not to equip yourself with radiation dosimetry equipment for a short trip to Tokyo, but if you intend to live there then you should do so by all means because Tokyo is quite badly contaminated. Likewise, if you are just passing through a city with high crime and murder rates, such as Chicago, it may not make sense for you to arm yourself or take self-defense courses in preparation, but if you intend to spend an extended period of time there, you probably should consider doing so.

Almost every one of us is exposed to a number of low-level but constant risks, and the fact that they are low on any given day should not be allowed to obscure the far more significant fact that their cumulative probability over time can approach 100 percent. Moreover, many risks are increasing over time. Extreme weather events—floods, hurricanes, heat waves—are constantly setting new records. Financial and economic systems are becoming increasingly unstable, causing people in various parts of the world to lose their incomes and their savings, and to become unable to buy what they need. Civil unrest, rioting and crime waves, such as the pandemic of sexual assaults on women by migrants currently gripping Sweden and Germany, are on the increase. The unmistakable overall global trend is toward greater instability.

Life in dense, built-up urban environments involves many additional risks. As a uniform background effect, wherever people (or, for that matter, animals) are confined in crowded conditions, they become skittish or aggressive. Crisis conditions bring this effect to the foreground. In big cities, society starts breaking down as soon as electricity goes off. In circumstances of great social inequality and injustice, where miserable populations live with repressed rage, we can expect widespread looting, assaults and general mayhem to erupt almost instantly. Another big area of risk is the complete dependence of urban populations on infrastructure services: just the failure of water supply or trash removal can make city life very unpleasant very quickly and almost impossible in a matter of weeks. Those whose lives are exposed to such chronic risks need to plan ahead, prepare and know what to do when a crisis hits.

There are three distinct kinds of plans that make sense, and you should work on developing each of them in parallel:

1. Prepare to remain in place and ride out the crisis. This requires you to create a stockpile of food, water, medications, other supplies such as cooking fuel, kerosene for lamps, candles, security measures and weapons for self-defense.

2. Prepare to evacuate. Have a bag packed with all of the essentials you will need in order to leave in a hurry: clothes, documents, cash; a reserve of fuel or an open ticket. Have an escape route worked out and a temporary destination where you intend to ride out the crisis.

3. Arrange a homestead in a rural setting or in a small town. It should be equipped to make it possible for you and your family to survive there for extended periods of time with minimal outside help. In addition to the usual short-term stockpiles of food and water, it should contain goods for barter (cigarettes, liquor, ammunition) and provide you with independent means of feeding yourself by growing food, gathering wild foods, hunting and fishing.

These three plans work well in combination. At the onset of a crisis, it is often hard to tell how long it is going to last, and until this becomes clear a reasonable plan of action may be to remain in place. In some cases, the crisis will be over before your stockpiles are depleted: floodwaters recede, electricity comes back on, law and order is restored, mountains of trash are hauled away, shops restock and reopen... But as the technosphere becomes increasingly starved of resources, the recovery period after each crisis is likely to grow longer and longer. At some point recovery will become impossible. (As ocean levels rise, seawalls and neighborhoods at low elevations along the coast, once they are damaged by a hurricane, may never be repaired at all.) If recovery appears to be taking too long to make it possible for you to remain in place, you may decide to evacuate to a safer location. Once there, you would need to carefully monitor the situation, eventually deciding either to return or to go somewhere else.

Clearly, the best place to resettle is your homestead. Here, the options are almost infinite. Your homestead can be a cottage, a farmhouse, a houseboat, a sailboat, a tiny house, camper, teepee or tent. It can even be a shipping container parked on a patch of land and packed with all the tools and supplies you need. All of these homesteads share two commonalities: they have to be set up ahead of time, and you need to practice living there before circumstances force you to. A crisis is a particularly bad time to try new things and to discover that some vital component is missing or doesn't work.

We have to accept it as an unfortunate fact that we will have almost zero control over the upheavals that will visit themselves upon us in the coming years and decades. Almost everything that the technosphere currently provides and that we take for granted— be it money, shopping, electricity, running water, trash removal, police protection, road transport, air travel, internet access, cell phone coverage, emergency services, government services, medical care—will at some point cease to function or become unavailable to us. What we can control is whether or not we prepare for these eventualities and how well we prepare.

It is to be expected that the vast majority of people will do almost nothing to prepare. When disaster strikes, they will run to the nearest shop and empty the shelves of anything useful. Then most of them will passively wait for help, complaining bitterly when that help is slow in arriving. Needless to say, that is not a good plan. It is easy for you to do better; it is somewhat harder but quite possible to do quite a lot better; and it is quite hard but still far from impossible for you to do perfectly well. In the process of preparing, you will transform yourself into a different kind of being—one that stands a good chance of living a reasonably satisfying life outside of the technosphere.

Stepping outside of yourself

IN MAKING THE transition, the first step is actually quite simple: you have to learn how to step outside of yourself. We are all creatures of habit, and our habits constrain us in what we can do. Most people go through life habitually doing just three things: doing as they are told, taking care of their needs and indulging their appetites. But in order to make the transition to autonomy, self-sufficiency and freedom we need to acquire one more habit—a meta-habit, if you will: the habit of breaking habits. It is the habit of being on the lookout for compulsive behaviors and suppressing them before the compulsion becomes hard to control. This meta-habit makes it possible for us to follow our own orders, acting in reasoned, decisive, self-motivated ways.

I have done so, and went on to break many habits, big and small. As a very minor matter, I used to smoke for quite a few years, but I no longer do. Nicotine is the most addictive substance there is, worse than heroin (so they tell us), and it took me a few tries before I finally quit. I didn't like the toll smoking was taking on my health and sense of well-being. I was disturbed by the sight of older people who were always hoarse, short of breath, coughed constantly—but still smoked. I did *not* want to end up like them. I hated having to waste money on tobacco. The fact that I still smoked made me feel weak-willed, and this lack of belief in my own willpower had a negative effect on other aspects of my life.

And then one day I had a realization: it is so much easier to *not* smoke than to smoke! If you smoke, you have to earn money, buy tobacco, roll it into cigarettes (I rolled my own to save money), light the cigarettes, suck on them, inhale, exhale, stub them out and do it all over again a short while later—all just to feel "normal."

All of that activity amounts to a lot of unnecessary work! Now, compare that to *not* smoking: you sit down, calmly fold your arms, close your eyes, and think: "Fuck you!"—the "you" in question being the old you—the smoker, the weak-willed coughing wreck, whom you obviously don't like or respect too much. All you have to do is psychologically disassociate yourself from your old self. Of course, you won't feel the least bit "normal" for quite a while, but you will know exactly why that is, and as with any pain or discomfort, if you know the reason for it and know that it isn't dangerous, then you can even learn to enjoy it! The pain from a bullet wound is bad pain; the pain from sore muscles after a long bike ride is good pain. And it turns out that pain from withdrawal symptoms can be transformed into good pain—if you put your mind to it.

And so it is with most habits: you stop being the person you don't like—one who lacks willpower—and you become a person who has more willpower and more self-respect. The older, unliked, weak-willed you still exists somewhere inside your skull but doesn't get to make any more decisions. The strangest realization I had about quitting smoking is that it is actually a useful exercise: becoming addicted to nicotine and then conquering that addiction provides a way to almost completely suppress the addictive part of your personality. Learning to break your most powerful habits and overcome your strongest urges makes it easier to conquer the rest of them. (Not that I would recommend that you start smoking for the sake of the exercise; I am sure there are plenty of other things for you to quit.)

Rites of passage

ONCE YOU CONQUER your compulsions, the next step is to conquer your fears and to gain self-confidence. I went through this process when my wife and I sold the house and car, and moved aboard a sailboat. I joyfully quit my corporate job and we sailed off. At that time, we hadn't had more than a daysail's worth of experience; the boat was old but new to us and had many minor problems of which I knew nothing. On our first trip out we sailed from Boston to Maine, which has a ragged coastline hiding a recently submerged mountain range. It is basically a giant pile of rocks with numerous submerged dangers and swift tidal currents. The weather is changeable and often fierce, and even in the summer it can be foggy and blowing a gale all at the same time.

The overnight passage from Boston to Portland was uneventful, but when we took off across Casco Bay and toward points further north it started blowing a gale as soon as the sun set. Next, the autopilot broke. Rather than attempting to thread the needle between rocky islands in the dark, I made the decision to head out into the open Atlantic. I trimmed the sails and lashed the tiller so that the boat went to windward without our having to steer it. By the time the gale died down we were quite far from any of the rocks. I took down the sails and slept. At dawn, we set course for a perfectly pleasant little harbor at Isle au Haut in Penobscot Bay, where we anchored and spent a few days hiking, picking wild mushrooms and swimming in the island's lake.

To an experienced sailor this should all sound quite humdrum: a short-lived summer gale in the North Atlantic, the autopilot dies (as they all do at some point), the skipper makes the right call to avoid land, sets a reasonable close-hauled course to achieve self-steering and rides it out. But to a seafaring newbie, which is what I was at the time, there was a short period of terror: I was tired, on a violently pitching boat in the dark, sails flogging, forced to make a swift decision. We were never in any sort of serious danger, but in my own mind I wasn't in a comfortable place.

I almost revisited that uncomfortable place a few thousand sea miles later, off Cape Hatteras, when we got caught in one of the spiral arms of Hurricane Bertha. We were a few days out of Florida heading up the coast, planning to make landfall in North Carolina, but not in the middle of a hurricane blowing onshore against the tide! The wind was howling, and once we got into shallower coastal waters (60 feet or less) the seas became quite ridiculously steep. Sure enough, the autopilot took this opportunity to act up again, resulting in an accidental gybe, which twisted the gooseneck off the mast, making it rather hard to continue sailing. When I tried the engine, it failed to start, making it also rather hard to continue under engine. The wind and the waves were pushing us toward the beach.

At that point, I made a decision, raised the Coast Guard on VHF channel 16 and asked to be towed through the inlet. As anyone who has been in this situation knows, once you contact the Coast Guard, you are pretty much just saying "roger" a lot and following their orders. It all worked out quite well: we raced toward the inlet under bare poles, where a tow boat passed me a line which, on third try, I secured to the bow. After that, the experience was quite similar to waterskiing. The tow boat captain opened up the throttles, and we literally planed through the inlet, green water sweeping across the deck and drenching us. Once inside the inlet, in perfectly calm water, I had to spend a good 15 minutes coiling all the line that had been swept off the deck or washed out of the cockpit and was trailing behind us, before we were sufficiently presentable to be towed to a marina.

I said that I *almost* revisited that uncomfortable place in my mind because I actually didn't. All it took the second time around was for me to order myself to shut up and just do my job. After that I was perfectly at peace with myself and the role into which I had been thrust and actually quite amused by all the giant mountains of water rising up and crashing all around us, with spindrift trailing off the crests, accompanied by flying balls of foam. I imagined that that's what an ant sees when looking up at a cabbage patch on a rainy, windy day. People at the marina to which we were towed had been monitoring all the back-and-forth on the VHF and were apparently quite impressed with the performance because they greeted us by handing us cans of beer as soon as we docked.

• • •

THERE IS A saying I happen to like:

Adventure is a sign of incompetence.

I find that there is a lot of truth to it; after all, consummate professionals don't go out looking for adventure and rarely find any. After a certain level of competence is reached, you are simply playing it safe all the time while executing your plans, and adventure becomes routine. After all, why would you take unnecessary risks if you know better? That is why the adventures of incompetent people make interesting stories, while the adventures of competent people do not—because they aren't really adventures at all but random mishaps.

This saying also implies that if you realize that you are incompetent but do not wish to remain so, you should perhaps go out and have some adventures. If you survive, not only will you gain some measure of competence (lessons learned in extreme circumstances are not soon forgotten) but you will also acquire certain other qualities—self-confidence, fearlessness, a certain inner calm—that will make it easier to face any other adversity in spite of your incompetence.

Taking this a step further, you can think of surviving adventures in spite of incompetence as a special skill. Given what our future holds, we are all going to be rendered incompetent at some point. Yes, we are all amateurs when it comes to surviving the future. Of course, there are still many things that are worth learning to increase your level of competence, but perhaps you should also work on becoming better at compensating for your incompetence—by having some adventures.

AFTERWORD

THE TRANSFORMATION FROM somebody who is spellbound by political technologies—to someone who is panicked and unsure—to somebody who is focused, resolved and engaged in forging a new path must involve some amount of pain. Reading this book may perhaps start this transformation, but completing it is up to you. The greatest fear to overcome, at least in my experience, is the fear of stepping outside of yourself and assuming another identity—the identity of someone who can just go ahead and deal with the situation.

As we grow up in civilized society, we are measured, evaluated, educated and trained, licensed and credentialed, and are eventually classified and assigned our place. In the process, we become very attached to this artificial identity that has been thrust upon us and very reluctant to set foot outside of it. (When we do, it is often when we are on vacation. It may therefore be helpful to think of this transition as going on a permanent vacation.) But if we overcome one fear—the fear of losing our assigned place in society—then many other fears fall away. Rational fears—ones that are based on an accurate perception of danger—do and should remain, but the irrational fear of stepping outside yourself and becoming someone else tends to disappear. And this opens us up to making dramatic changes, adapting to new circumstances and environments and, in the process, setting ourselves free.

INDEX

C

cancer, 37, 45
   rates of, 50–51, 78, 112
capitalism, 2, 80, 90
Cargill company, 113
Chernobyl nuclear disaster, 112, 226, 235
children
   and future, 3, 57, 107, 147, 151, 209,
      215, 231–32, 235
   and learning, 13, 221–22
   as self-reliant/independent-minded,
      67, 147, 223
   as simulated, 32–33
   in weakened state, 22, 31, 40, 68, 78,
      115, 166, 169–70
Chomsky, Professor Noam, 45–46
Christianity
   as form of technology, 82
clathrates, oceanic, 46
climate change, 153, 161–62, 209, 227, 234
   as disruption, 38, 56, 158, 226
   politics of, 210
climate science, 210
Clugston, Christopher, 74
coal, 15, 43, 74, 129, 225
Color Revolution, 127, 170–72, 175, 177–79,
   181–83
commando units, 183
communication, face-to-face, 180
communication networks, 55
communications, electronic
   progression of, 30
communications technology, 192
competence
   in surviving future, 244
computer algorithms, 16, 59, 67
computers, wearable, 32
computer simulations, 33–34
cone tent, 157
Confessions of an Economic Hit Man, 169
control, total. *See* technosphere, total
   control
corporate profits, 2, 5, 114
corruption
   in government and banking, 52
creative professions
   as marginalized, 19
crop failure, 13

crowd control methods, 5
crude oil, 41, 43–44, 47, 73
Cultural Revolution, 24
culture, 22–24, 61, 63–65, 79, 104, 129, 132
   nomadic, 155–56

D

death, 57, 68, 94–95, 107, 111, 186
   as technical limitation, 17–18
decision-making, 19, 192
   by technology, 58–59, 68, 191
Deepwater Horizon disaster, 56
Deer Island Sewage Treatment Plant, 82
deindustrialization, 95
dementia, 8, 33
democracy, 63, 93, 163, 181–82, 209, 226
dependency, 22, 28, 99, 104
   on machines, 4, 212
   on technology, 6, 91, 111
   on technosphere, 13, 40, 59, 72, 202,
      215, 237
dictatorship of the law, 127–28
diesel fuel, 44, 103, 122
digital information, 103, 219
direct democracy, 181
disaster preparations, 230, 232–33, 235,
   238–39
disequilibrium, 54
distractions
   from shrinking the technosphere,
      144, 178, 207–10
diversity, 10, 22, 53, 60, 65
dome tent, 157
Dunbar, Robin, 211

E

Earth, 15, 21, 39, 47, 49, 53, 77, 129, 155
   and climate, 56, 144, 158
economic development, 6, 30, 43, 69
   as global, 25, 170
   as stalling out, 227
economy of personal means, 7
educational standards, 30
electroconvulsive therapy, 27
Ellul, Jacques, 81–86, 90
emergent intelligence, 53, 80, 225
   as machine-like, 117
enclosure movement, 24

human genome, 44
human labor, 2, 15–16
human rights, 4, 63, 226
hunting, of food, 40, 61, 76, 216, 238
hyperalienation, 195

I
ideology
        to unite partisans, 186–87
IMF (International Monetary Fund), 184
immune systems, 109
industrial agriculture, 53
industrialization, 52, 58, 137
Industrial Society and Its Future, 89
industrial technology, 81, 108, 130
infectious diseases, 57, 234
information technology, 116, 191
innovations, 7, 12, 35, 91, 157
instability, 76, 112, 177, 227–28, 237
intellectual labor, 16
intellectual property laws/rights, 101,
        219–20
Intergovernmental Panel on Climate
        Change (IPCC), 152
International Loan Sharking, 169, 177–78
internet access, 4, 31, 98, 106, 136, 191,
        207, 221
internet addiction, 3, 31
"internet of things," 131
iron triangle, 203, 205
Islamic extremism, 29

K
Kaczynski, Ted, 86–96, 98
Karelo-Finnish Soviet Socialist
        Republic, 133
Kasparov, Gary, 127
Kerrigan, Sean, 196
Kessler syndrome, 39
Khodorkovsky, Mikhail, 127
killing technologies, 48
Koryak-Chukchi *yaranga*, 157
Kyoto Protocol, 210

L
Laughlin, Dr. Greg, 39–40
leftism, 91–92
legal system, 167
Lenin, Vladimir, 185

liberalism, 63
life, extension of, 18
lifestyle
        and harm/benefit analysis, 125
        migratory, 13, 62, 77, 154
        nomadic, 13, 62, 77, 154–55, 157
        settled, 154–55, 158
lithium, 74–75, 121–22, 219
live-aboard (boats) lifestyles, 216–18
living things, 51, 54, 66, 94, 116
        and science, 27, 45
        uncontrollable messiness of, 26, 46
Lovelock, James, 10, 53

M
machine labor, 2, 15
Mars Climate Orbiter, 39
mass media, Western-controlled, 173,
        175, 179, 183
medical system, 17, 71–72, 165
methamphetamine, 23
microorganisms, 26, 55
mind control methods, 5
minorities, sexual, 29
Minsk-I & II, 185
mixed oxide (MOX) fuel, 36
money, 111, 216–17
        as counterflow, 103
        as fungible commodity, 21–22
        need for, 208, 214
        as value of everything, 20
monoculture, 26, 53, 57
Monsanto company, 113, 115, 119, 166
morality
        as controlling humans, 29
Mubarak, Hosni, 171, 179–80
multiculturalism, 22, 24, 65
Mumford, Lewis, 190
murders, 76, 169, 237
        high rate of, 70
        as officially sanctioned, 69

N
nanotechnologies, 49, 129
        unlimited harm of, 115, 119
NASA Ames Research Center, 39
national defense establishment, 164
national security, 30
NATO, 78, 185, 229

ABOUT THE AUTHOR

DMITRY ORLOV WAS born in Leningrad, USSR, into an academic family and emigrated to the US in the mid-1970s. He holds degrees in computer engineering and linguistics, and has worked in a variety of fields, including high-energy physics, internet commerce, network security and advertising.

Since 2005, Dmitry has published hundreds of articles, two books and five books of essays. He has given numerous talks and interviews, and delivered keynote addresses at many conferences. His work has been translated into many languages.

A decade ago Dmitry made a dramatic change in lifestyle, trading dependency and financial security for resilience, self-sufficiency and freedom. He gave up on corporate employment in Boston's high-tech sector, sold the condo and the car, bought a sailboat and set off sailing. This experiment has yielded a wide variety of insights into just how far it is possible to downscale and simplify one's lifestyle while remaining productive, comfortable and civilized; which skills and technologies are needed; and which are superfluous. Having to decide which specific elements of technology are appropriate to this lifestyle, which are not, and which are harmful naturally caused him to focus on the wider problem of making conscious and deliberate technological choices.

A Note About the Publisher

NEW SOCIETY PUBLISHERS is an activist, solutions-oriented publisher focused on publishing books for a world of change. Our books offer tips, tools and insights from leading experts in sustainable building, homesteading, climate change, environment, conscientious commerce, renewable energy and more—positive solutions for troubled times.

Sustainable Practices for Strong, Resilient Communities

We print all of our books and catalogues on 100% post-consumer recycled paper, processed chlorine-free, and printed with vegetable-based, low-VOC inks. These practices are measured through an Environmental Benefits statement (see below). We are committed to printing all of our books and catalogues in North America, not overseas. We also work to reduce our carbon footprint, and purchase carbon offsets based on an annual audit to ensure carbon neutrality.

Employee Trust and a Certified B Corp

In addition to an innovative employee shareholder agreement, we have also achieved B Corporation certification. We care deeply about what we publish—our overall list continues to be widely admired and respected for its timeliness and quality—but also about how we do business.

New Society Publishers
ENVIRONMENTAL BENEFITS STATEMENT
For every 5,000 books printed, New Society saves the following resources:[1]

30	Trees
2,695	Pounds of Solid Waste
2,965	Gallons of Water
3,867	Kilowatt Hours of Electricity
4,899	Pounds of Greenhouse Gases
12	Pounds of HAPs, VOCs, and AOX Combined
7	Cubic Yards of Landfill Space

---

[1] Environmental benefits are calculated based on research done by the Environmental Defense Fund and other members of the Paper Task Force who study the environmental impacts of the paper industry.